A Python Primer for ArcGIS®

Nathan Jennings

Copyright © 2011 Nathan Jennings
All rights reserved.
ISBN: 146627459X
ISBN-13: 978-1466274594

ACKNOWLEDGEMENTS 11

INTRODUCTION 13
    Objectives and Goals 13
    Prerequisite Knowledge and Skill 15
    Problem Solving 15
    Developing Geoprocessing Workflows 17
    Structure of the Book 20
    Data and Demos 21
    Obtaining the Supplemental Information 22
    Required Software 22
    Older Versions of ArcGIS and Python 23
    Author's Note 23

SECTION I: THE FUNDAMENTALS 25

Chapter 1 Python and ArcGIS 27
    Overview 27
    Python and ArcGIS Versions 29
    How Python is used with ArcGIS 31
    Python Development Environments 31
    Relationship to ModelBuilder 34
    Python Shell in ArcGIS 34
    Use of Scripts with Geoprocessing Tools 37
    Getting Help 37
        ArcGIS 38
        Python 39
    Python and ArGIS Errors 39
        Python Syntax Errors 40
        ArcGIS Error Codes 41
    Common Methods for Handling Errors 44

Chapter 2 ModelBuilder and Python 47
    Overview of ModelBuilder 47
    ModelBuilder Python Script Caveats 48

Demos Chapter 2 51

Demo 2a: Using ModelBuilder to Create a Python Script 52

Demo 2b: Build the Clip Model Two Ways and Export to Python Script 63

Exercise 2: Create a New Feature Class and Add Fields 70

Chapter 2: Questions 71

Chapter 3   Python Constructs 73
Overview 73
Using Python IDLE for Code Development 73
Using the Python Shell for Code Testing 76
Syntax 77
   Case Sensitivity 78
   Naming Conventions 78
   Indentation 79
   Comments 80
   Creating and Using Variables 81
   String and Number Variables 82
   Lists 83
   Conditional Statements and Loops 85
import Modules 86
try: and except: Blocks 87
Special Considerations for Strings in Python 88
   Backslash, Forward Slash, and Raw Strings 88
   Single and Double Quotes 89
General Structure of a Python Script 93
   Title, Author, Date, and Script Comments 95
   import Modules 95
   Variable Definitions (and Python function definitions) 95
   Code Body 96
Running a Python Script 96
   Check Module 96
   Run Module 98
   Handling Errors 99
Summary 101

Exercise 3: Write a Simple Python Script 102

Chapter 3: Questions 105

Chapter 4   Writing a Basic Geoprocessing Python Script 107
Overview 107
Getting Ready to Create an ArcGIS Geoprocessing Python Script 107
Using Pseudo-code to Outline Geoprocessing Tasks 108
arcpy Module Overview 110
Workspace and Data Path Variables 113
Define Variables 115
Hard Coded Parameters 115

Parameters Using Variables 116
Add and Modify Geoprocessing Functions 116
Search ArcGIS Help 117
ArcGIS Toolbox Aliases 121
Summary 122

Demo 4: Writing a Clip Features Script 123

Exercise 4: Add the Buffer Routine to the Clip Features Script 134

Chapter 4: Questions 137

SECTION II: WRITING PYTHON SCRIPTS FOR COMMON GEOPROCESSING TASKS 139

Chapter 5 Querying and Selecting Data 141
Overview 141
Prerequisites 145
Building the Query Syntax 145
    Field Name Syntax 146
    Developing and Processing Strings in Query Expressions 147
    Common Operators in Queries 151
    Wildcard Characters 152
    NULL Values 152
    Numerical Expressions in Queries 153
    Using Calculations in Queries 154
    Combining Expressions 154
Other Query Syntax 155
Feature Layers and Table Views 156
    Make a Feature Layer 159
    Make a Table View 160
Selecting Data 161
    Select Data by Attribute 161
    Programming the SelectLayerByAttribute Routine 165
    Select Data by Location 166
    Programming the SelectLayerByLocation Routine 169
Counting the Number of Records 171
Creating a New Dataset 171
Data Locks 172
Summary 176

Demos Chapter 5 177

Demo 5a: Create a Feature Layer with and without a Query 178

Demo 5b: Select Features by Attribute 185

Exercise 5: Select Features by Attribute/Location and Write them to a New Feature Class.... 202

Chapter 5: Questions **204**

Chapter 6   Creating and Using Cursors and Table Joins 205
Overview 205
Cursors 205
Types of Cursors 209
Implementing Cursors 209
    Search Cursor 210
    Creating and Using the Search Cursor 210
    Using the for Loop 212
    Using the while Loop 214
    Insert Cursor 216
    Schema Locks on Data 218
    Creating and Using the Insert Cursor 220
    Update Cursor 223
    Creating and Using the Update Cursor 225
Table Joins 230
Attribute Indexes 234
Programming and Using Table Joins 238
    Create an Attribute Index 238
    Create Feature Layers or Table Views 239
    Create the Table Join 240
    Using and Accessing Information in Joined Data 240
Summary 243

Demos Chapter 6 .... **244**

Demo 6a: Search Cursor 245

Demo 6b: Insert Cursor 250

Demo 6c: Search and Update Cursor 258

Demo 6d: Joining Tables 266

Exercise 6: Using Cursors and Table Joins 273

Chapter 6: Questions **278**

Chapter 7   Describing Data and Operating on Lists 281
Overview 281

Describing Data 281
Listing Data 286
Summary 290

Demos Chapter 7 .... **291**

Demo 7a: Describe Properties of an Image **292**

Demo 7b: Listing Data **296**

Exercise 7: Batch Clip Images Using a Feature Class **300**

Chapter 7: Questions **304**

Chapter 8   Custom Error Handling and Creating Log Files **305**
Overview 305
Custom Error Handlers 305
Creating and Using an Error Handling Class 306
Using Log Files to Collect Messages 310
   Creating and Using a Log File 310
   Adding the Date or Time to a File Name or Message 312
Summary 313

Demo 8: Create Custom Error Messages **314**

Exercise 8: Create Custom Error Messages and Log File for a Script **318**

Chapter 8: Questions **319**

Chapter 9   Mapping Module **321**
Overview 322
Map Elements 323
Map Element Relationships 327
Prerequisites 328
   Create a Map Template 329
   Name Map Layout Elements 329
Mapping Module Class Properties and Methods 332
Mapping Module Functions 333
Implementing Map Documents, Data Frames, Layers, and Layout Elements 333
   Import the `mapping` Module 334
   Accessing an ArcMap Document 334
   Accessing a Data Frame 336
   Accessing Layers 338
   Using Layer Properties in the Script 340
   Accessing and Changing Layout Elements 342

Exporting Map to PDF or Print Map to Printer 344
Saving Map Documents 346
Working with Data Frame, Layer, and Layout Element Methods 347
    Changing the Map Extent using a Definition Query (or not).. 347
    Changing the Map Extent using Selected Features 349
    Adding and Saving Layer Files 353
Creating a Map Book Programmatically using the `mapping` *Module* 356
Summary 357

Demos Chapter 9 359

Demo 9a: Mapping Module Overview and Properties 360

Demo 9b: Implementing Mapping Module Methods 368

Exercise 9: Create a Simple Neighborhood Map Set 376

Chapter 9: Questions 378

SECTION III: INTEGRATING AND AUTOMATING PYTHON SCRIPTS FOR ARCGIS 381

Chapter 10   Custom ArcGIS Tools and Python Scripts 383
Overview 383
Creating a Custom ArcToolbox 386
Associating a Python Script to the Custom Toolbox 388
Defining Parameters for the Script Tool 391
    Parameter Display Name 394
    Parameter Data Types 394
    Parameter Properties 394
Customizing the Script Tool Interface 400
Modifying the Python Script 401
    Adding Python Script Parameter Arguments 401
    Adding ArcGIS Messages to the Python Script 405
Executing the Script Tool 407
Writing Tool Documentation 410
Summary 417

Demo 10: Create a Custom Script Tool Interface for the Clip and Buffer Tool 419

Exercise 10: Create a Custom Script Tool Interface for Your Own Script 428

Chapter 10: Questions 429

Chapter 11   Automating Geoprocessing Scripts 431
Overview 431
The Batch File 432
Running a Python Script at a Command Prompt 432
Creating a Python Batch File 437
Scheduling the Batch File to Automatically Run the Geoprocessing Script 439
Summary 446

Demo 11: Create and Schedule a Batch File to Auto-run a Python Script 447

Chapter 11: Questions 449

APPENDICES 451

Appendix 5.1   Python, Cursors, and Open Source Databases 451

Appendix 9.1   Summary of the mapping Module Properties, Methods, and Functions 452

REFERENCES 458

INDEX 459

Acknowledgements

A Python Primer for ArcGIS® is a culmination of the author's experiences and relationships with a number of people and organizations and could not have been written without them. The author would like to acknowledge the Environmental Systems Research Institute (Esri®), the company that provides geographic information systems (GIS) software to most of the world's GIS users. This organization and software has made it possible for many people and organizations to explore, analyze, and depict their world using geographic information. Specific to this book, Esri has developed modules and objects that can be used with the open source Python programming language. Doing so has allowed their software to become more customized and expanded for specific geoprocessing tasks.

The author would also like to acknowledge the City of Sacramento and ICF International (formerly, Jones and Stokes). These organizations provided the impetus for the author to develop his own Python programming skills and knowledge and are sources for some of the demonstrations and exercises in this book. In addition, the author would like to acknowledge American River College in Sacramento, CA, the geography and science department, and especially the students in the GIS Program. The author developed and teaches the on-line GIS Programming course at American River College and the students have served as the "testers" of the material in this book. Their feedback has been valuable for many of the edits that went into this book.

The author acknowledges the full GIS staff at the City of Sacramento. These colleagues have been some of the best to work with over the author's career and represent some of the finest GIS professionals in the community. Specifically, the author would like to mention Dan McCoy. Dan has been a valuable resource to bounce ideas off of and to help clarify some of the coding logic and geoprocesses that found its way into the text. In addition, the author would like to thank the Central GIS team that the author works with. In addition to Dan, the team includes Maria MacGunigal, David Wilcox, Rong Liu, and Carlos Porras. The author would like to especially thank Dr. Este Geraghty who took her time as a student in the author's class and with her very busy schedule to review, comment, and make suggestions for *A Python Primer for ArcGIS*. Her feedback is sincerely appreciated.

Cover design by Zach Jennings; web support by Josh Jennings.

Esri® ArcGIS® software graphical user interfaces, icons/buttons, splash screens, dialog boxes, artwork, emblems, and associated materials are the intellectual property of Esri and are reproduced herein by permission. Copyright © 2011 Esri. All rights reserved. Esri, ArcGIS, ArcInfo, ArcEditor, ArcMap, ArcCatalog, ArcView, ArcSDE, ArcToolbox, 3D Analyst, ModelBuilder, ArcPy, ArcGlobe, ArcScene, ArcUser, and www.Esri.com are trademarks or registered trademarks or service marks of Esri and are used herein by permission.

Introduction

For the last several years Esri has supported the use of the open source scripting language Python for many of its geoprocessing tools and functions within ArcGIS. Python is platform independent, so it serves as a good single common scripting language for different operating systems as well as for different versions of ArcGIS. As ArcGIS development moves forward, organizations and individuals will not need to maintain geoprocesses using multiple scripting languages such as Arc Macro Language (AML™) and Avenue™, which are now both outdated and not officially supported.

Python is rapidly becoming the scripting language that geographic information systems (GIS) professionals and newcomers to GIS want and need to learn. They see knowing how to program, especially using Python will be beneficial to their career and help existing organizations that have a long history of scripting development to transition to more current geoprocessing standards, especially if the organization uses Esri software, which many do.

Objectives and Goals

This book is written for those who want an introduction to Python using ArcGIS geoprocesses. A *Python Primer for ArcGIS* is not a detailed text on Python. Others have already accomplished this task. References can be found throughout and at the end of the book. *A Python Primer for ArcGIS* will help newcomers to GIS and programming as well as those who are strong users of ArcGIS geoprocesses but do not yet have a solid knowledge or expertise in writing scripts. For those who have some background in programming, many of the concepts such as variables, loops, conditional statements, etc will be familiar and helpful in developing Python scripts. For those who do not, this book will serve as a starting point to begin developing code using some of the basic programming structures that are commonly used in many of the geoprocessing tasks found and used within ArcGIS. *A Python Primer for ArcGIS* focuses on developing geoprocesses and Python script in a logical manner to primarily develop standalone scripts that can be run both inside and outside of ArcGIS. A *Python Primer for ArcGIS* accomplishes the following objectives within this book:

1. Provide a framework for code developers at different levels to design logical geoprocesses as well as design logical coding structures that include proper constructs for error handling, troubleshooting processes, logic, and scripting problems
2. Introduce common Python constructs and how they are implemented with ArcGIS geoprocessing tools
3. Assist the code developer to obtain help with Python, ArcGIS geoprocessing functions, while building their own code writing skill
4. Introduce some of the new functionality of Python and ArcGIS, such as the mapping and spatial analysis modules
5. Show how to integrate custom built scripts with the ArcToolbox™
6. Show how to auto-run custom scripts

The primary goal of *A Python Primer for ArcGIS* is to provide the user an organized path to gain a solid understanding of the common Python elements used in ArcGIS as well as to demonstrate how Python can be used for the most widely used geoprocessing tasks. These tasks will make up the majority of the book's content and is the primary reason why the author focuses on developing standalone scripts. With a grounding in both Python structure and syntax and common ArcGIS functions, the reader will be able to apply this knowledge, skill, and ability to more complex scripting and geoprocessing tasks (e.g. Python dictionaries, arrays, functions or ArcGIS extensions, ArcSDE®, and specialized geoprocessing methods) while having a structured environment to develop geoprocessing and scripting workflows. Upon reading and studying the concepts in *A Python Primer for ArcGIS* and performing the demonstrations and exercises and answering the chapter questions, the reader should be able to design, develop, create, troubleshoot and successfully run Python scripts with multiple steps and multiple ArcGIS geoprocessing functions and methods.

Prerequisite Knowledge and Skill

The reader and user of *A Python Primer for ArcGIS* should have a fundamental understanding of GIS concepts such as geographic features (points, lines, and polygons), feature classes, GIS geospatial data formats, data and attribute tables, relational databases, records, rows, fields, columns, etc. as well as a fundamental understanding of ArcGIS, how it is structured, and how to use ArcMap™, ArcCatalog™, and ArcToolbox™. The reader and user should also know how to use some of the geoprocessing tools (Clip, Buffer, Select Layer by Attribute, Select Layer by Location, etc.) within ArcToolbox. Familiarity with ModelBuilder™ is recommended, but not required to use this book. Completing some of the Esri courses or similar introductory college GIS courses that use ArcGIS should be able to provide the requisite knowledge to get started with *A Python Primer for ArcGIS*.

The reader does not need to know how to program or know Python or any other programming language. This text will provide an introduction to Python and general Python programming constructs that can be used with ArcGIS geoprocessing functions.

Problem Solving

Problem solving is an important skill to develop in an analytical field such as GIS. As a GIS professional and college instructor the author has developed a variety of problem solving skills that he uses every day in his work, uses and communicates with colleagues and clients, as well as teach students in the classroom. These skills may take a variety of forms:

1. Determine and develop which GIS tasks are needed to implement a series of geoprocessing analyses
2. Determine which systems integration and data management tasks are required to generate effective and efficient operational business workflows
3. Develop which programming tasks and the order of the geoprocessing tasks are required to automatically process GIS data using Python

Developing problem solving skill is not easy and takes time, practice, and often takes hours of research to create solutions to GIS problems and scripts. The author spends considerable time reading and studying documentation, trying out specific geoprocessing functions, analyzing data and reviewing and interpreting intermediate and final results, develop and test specific workflows, and building simple to complex geoprocesses. One can think of this as a modified "Scientific Method." Readers are encouraged to consult ArcGIS help, on-line forums, study other developer's code, and build a repository of scripts and samples for future reference. From a code development point of view, many of the above techniques are used, since the proper result cannot be achieved without writing the proper code (instructions) for the "computer" to implement the script.

In addition, the author often creates written documentation (outside of in-line code documentation) that explain processes, methods, data input/output, and solutions to intermediate problems in "plain English" that can often be referred to as well as develop for more comprehensive and formal documentation that the author provides for others to use. The author encourages the reader to do the same. For those who enroll in the GIS classes at American River College, the author provides the opportunity to learn and develop problem solving skills. For those of who refer to this book, consult the sources above to develop your own problem solving techniques or contact the author for more information.

Developing Geoprocessing Workflows

Before a GIS person (or team) undertakes a geoprocessing problem, often a result, goal, product, or service is needed, desired, required, etc. These can take the form of creating a new data set, summarizing data to help make a decision, generating a set of maps to show results of geospatial analyses, providing a web service, or developing a process to manage and update data for a specific purpose. All of these tasks require some set of steps to generate the result and often require some kind of interpretation, analysis, and evaluation of data, intermediate results, and final results. The author has found it beneficial to develop a geoprocessing workflow *before* a project starts to outline or map out the data requirements, processing steps, intermediate results, and final results. *(Oftentimes, in practice, the workflow is never developed or only developed after realizing the hard work a person or team has expended).* Developing an outline or workflow before a project commences the GIS technician, analyst, or programmer can operate with a structured framework (i.e. the overall objectives and goals of the project). Having this larger perspective on a project or task, solid solutions, geoprocessing tasks, products, and services can be developed and be developed for the right purposes. In addition, an outline or workflow diagram provides documentation to the process that can be referred to, since many projects can take a number of weeks or months and a single person will likely not remember all of the specific tasks, data, and products, plus the GIS person will likely be working on multiple projects at any given time. These outlines and workflows can also be used in documents provided to other staff members, clients, or for internal documentation.

The workflow can take many forms such as an outline of steps or a workflow diagram indicating the relationships between one step and another or how one step may be related to many steps. For example, one source dataset may be used in multiple geoprocesses. This kind of workflow is often seen when designing a geoprocessing model in ModelBuilder. The workflow can be fairly simple involving a small number of geoprocessing tasks, or it can be complex involving many data sources, processing steps, feedback (looping) mechanisms, and many outputs (geospatial data, tables, maps, web services, etc).

The following figures show examples of actual outlines and workflows developed by the author for specific tasks and projects. The figure below shows an actual Python script created by the author that processes an attribute domain for a feature class. Notice the comments (marked with a # sign) that briefly describe the specific geoprocessing task. The commented steps (sometimes referred to as "pseudo-code") provide the framework for the script. The comments were actually written first before any specific ArcGIS geoprocesses were created so that the general set of steps could be outlined and the author could determine if other geoprocessing steps were required and could be researched for syntax, parameter, data, and data type requirements. The outline was modified as required during the development of the script.

```
gp.CreateTable_management(gp.workspace, sorted_parts_table, parts_table)

# 1. Access a table using a search cursor and the A identifier
#    to sort the data in ascending order based on the Code attribute

srows = gp.SearchCursor(parts_table, "", "", "", 'Code A')

irows = gp.InsertCursor(sorted_parts_table)

srow = srows.next()

# 2. Update the "sorted_parts_table" with the sorted records from the
#    existing table

print "Sorting Parts Table..."
print >> log, "Sorting Parts Table..."

while srow:

    irow = irows.NewRow()
    irow.Code = srow.Code
    irow.Description = srow.Description

    irows.InsertRow(irow)

    srow = srows.next()


# 3. Remove Existing Attribute Domain from field, in this case
#    each of the 5 sign attributes


print "Updating Domain for " + signs_fc + "..."
print >> log, "Updating Domain for " + signs_fc + "..."
x = 1
while x <= 5:

    gp.RemoveDomainFromField(signs_fc, 'FACE' + str(x))
    x = x + 1
```

As another example, the figure below shows a workflow diagram that was developed for a geoprocessing workflow to perform data maintenance on traffic signs for the City of Sacramento. This workflow is used by GIS staff in the city's department of transportation. A GIS data management document accompanies the workflow that provide specific GIS data processing tasks that the GIS staff uses to update and manage the city's traffic sign inventory. The GIS staff at the city uses this workflow to discuss similar activities for other city departments. A full discussion of the workflow and geoprocess implementation can be found in the Winter 2009 issue of *ArcUser*™
(*http://www.Esri.com/news/arcuser/0109/streetsigns.html*).

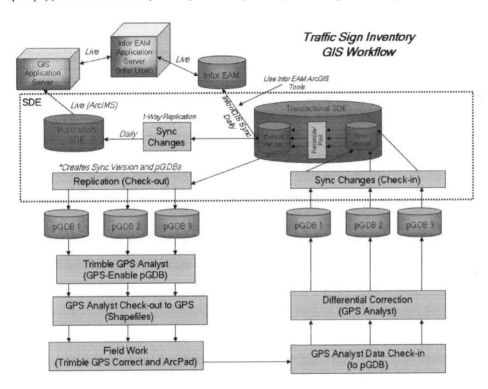

In both cases, a general process is outlined and documentation is created that can be used for a wide variety of purposes.

Structure of the Book

The structure of this book begins with a description of Python and ArcGIS fundamentals. The first chapters of the book discuss how Python is organized and structured and some of the fundamental requirements in order to develop code that is useful for ArcGIS. ModelBuilder is mentioned and how it can be used to develop geoprocessing logic and test ideas before exporting the work to Python for further development. In addition, the subject of "getting help" is discussed, since additional assistance with ArcGIS routines, syntax, properties, and methods will likely be needed during code development for both Python structures and ArcGIS geoprocesses.

After a brief overview of some of the common Python elements and the ArcGIS Tools in Section I, Section II of the book focuses on how to develop Python code for many of the commonly used GIS tasks. These include such topics as querying and selecting data, creating data, relating data, working on subsets of data, and creating new output. This section also includes the implementation of the mapping module with ArcGIS for automating map creation tasks. Section III focuses on integrating and automating Python scripts within the ArcGIS Toolbox or as back-end scheduled processes.

Most chapters will have a demonstration program that the reader can work on and develop using step by step examples. In addition, the reader can work on the chapter exercises to obtain more experience with developing code and to test the knowledge gained in the chapter and demonstration program. Most chapters have questions that focus on the important concepts.

The author uses the following typeface conventions throughout the book:

Street_CL – bold type typically indicates a feature class or table explicitly used in the text, demo, or exercise. Bold is also used to highlight ArcGIS Help documentation topics so the reader can easily find additional information provided by Esri.

StreetName – italics type typically indicates an attribute field. It will also be used to indicate a published work.

`arcpy.SearchCursor()` – Courier indicates example Python syntax within the text, demos, and exercises.

<required_parameter> - will indicate a required parameter for an ArcGIS tool or routine
{optional_parameter} – will indicate an optional parameter for an ArcGIS tool or routine

Data and Demos

All of the data and demo scripts can be found at the author's website at www.jenningsplanet.com\PythonPrimer. The supplemental material will be organized in the following manner: **\PythonPrimer\ChapterXX**. Within each chapter the **Data** folder contains the data files required for the demo and/or exercise. Data files can be shapefiles, file, or personal geodatabase feature class or tables, or standalone tables (e.g. dBase format). ArcMap documents (.MXD) can be used as referenced or renamed for readers to modify and save their own work. A **MyData** folder is also provided so that readers can save their own work for demos and exercises. All of the data and ArcMap documents will be in ArcGIS 10 format. NOTE: The ArcMap documents reference the **\PythonPrimer\ChapterXX** structure above. If the reader changes this folder structure, the ArcMap documents provided by the author may need to have the source files in the Table of Contents revised to the new location. The scripts have been tested on Windows XP and Windows 7 32-bit operating systems. The reader may need to make some additional adjustments to data paths on 64-bit Windows systems.

The data sources exist on the one of the following web sites or organizations:

City of Sacramento – city related vector data and historical 1991 aerial photos
County of Sacramento – parcel and street subsets
CalAtlas – Landsat Thematic Mapper (TM) satellite imagery subset

Refer to the text file associated with the supplemental data for more information as well as the websites at the end of the book.

Obtaining the Supplemental Information

The reader can obtain the supplemental information for this book at the author's website using the following credentials (and are case sensitive):

www.jenningsplanet.com/pythonprimer

Username: PythonPrimer
Password: PP4AGIS!

Additional information will be provided on this website with any updates, changes, etc.

Required Software

The user must have access to an ArcView®, ArcEditor™, or ArcInfo® license, version 10 and Python 2.6.5, the supported version of Python for ArcGIS 10. Students enrolled in the online Introduction to GIS Programming course (Geog 375) at American River College (http://wserver.arc.losrios.edu/~earthscience/) can obtain a one-year student license of ArcInfo 10. Contact the author to check enrollment and validate academic status.

Alternatively, the reader can obtain a copy of ArcGIS for Home Use at http://www.Esri.com/arcgis-for-home/index.html or from one of the ArcGIS books from Esri that comes with a CD and DVD. The CD contains the data, demos, exercises, and solutions; the DVD contains a 180 day fully functional copy of ArcView 10. Esri can be contacted to receive an evaluation copy of ArcGIS that can be used with this book. Readers with access to ArcGIS only need to copy the data referenced in the book to get started with *A Python Primer for ArcGIS*. Readers are encouraged to review their own data or a company's data collection and practice writing additional scripts beyond the exercises and demonstrations provided in this text.

Older Versions of ArcGIS and Python

Older versions of ArcGIS and Python can be used with this book, but the syntax and structure of some programming elements may differ considerably. The ArcMap documents (MXDs), files, and geodatabases will need to be opened with ArcGIS 10 and exported or saved to an older ArcGIS version. It is recommended that the latest version of ArcGIS be used with the content in this book. See the Author's Note below and Chapter 1 for more details on ArcGIS and Python compatibility.

Author's Note

As of the writing of this writing of this book, the current version of ArcGIS is ArcGIS 10 which uses Python 2.6. A number of readers will still be using ArcGIS 9.3.x and possibly 9.2 in their organization. This book focuses on the use of the *arcpy*™ module using ArcGIS 10 and Python 2.6. For code developers that require a Python script for use in ArcGIS 9.3.x or ArcGIS 9.2, Python 2.5 will be used with ArcGIS 9.3.x and Python 2.4 for ArcGIS 9.2, both of which use the *arcgisscripting* Python module and does not contain the `mapping` module. Script syntax will need to be written to support both the ArcGIS and Python versions. Code developers can consult the respective ArcGIS Resource Center web help http://resources.arcgis.com/content/web-based-help and can also go to www.jenningsplanet.com for some common sample scripts using previous versions of the *arcgisscripting* Python module.

Section I: The Fundamentals

Section I introduces the Python scripting language and how it relates to ArcGIS. Chapter 1 focuses on how Python can be used with ArcGIS and its relationship with ModelBuilder, since some readers may already have experience with ModelBuilder. Chapter 1 also introduces the Python script Interactive Development Environment (IDE), called IDLE so that the reader has a basic understanding of where to write actual Python script. The reader is provided a high level overview of how to obtain help with both Python and ArcGIS as well identifying errors that are likely to occur when developing Python code.

Chapter 2 focuses on ModelBuilder and provides a general overview of how ModelBuilder operates and how it can be used to develop geoprocessing logic and can ultimately be exported to Python script where it can be more fully developed.

Chapter 3 reviews the primary Python constructs that will be used throughout *A Python Primer for ArcGIS*. This chapter introduces these concepts at a broad level where many of them will be more fully discussed in the context of their use with ArcGIS geoprocessing tasks and functions.

At the end of Chapters 3 and 4, the reader has a chance to write a simple Python script as well as a simple geoprocessing script that will serve as the launch point to develop more complex geoprocessing routines using Python and ArcGIS.

Chapter 1   Python and ArcGIS

Overview

One of ArcGIS's primary functions is to perform geoprocessing and data management operations. These can include overlays (intersect, union, identity, spatial join), querying data (select records by attribute queries or select by spatial coincidence), and dozens of other geoprocessing functions such as those found in the ArcGIS Toolbox. ArcGIS can also be used to create and manage data in a variety of formats. These can include, create feature classes and tables, calculate attribute fields, update feature classes and attribute values, work with imagery and image processing tasks, and join data tables among others.

Python, an open source, cross operating system platform scripting language, provides a means to perform many of the operations mentioned above as automated tasks and batch processes. Instead of a user having to manually select multiple tools, fill in parameters, and clicking the OK button, analysts can write Python scripts to automate workflow tasks that can be run by an analyst either through the Python script interface or through a custom tool in the ArcToolbox. Python scripts can also be executed to run at a scheduled time, so that the automated tasks can run after business hours or during low network impact.

The ArcGIS help makes specific reference to two different uses of Python scripts:

1. Python Script Tool – this is a script that is written with the intent to be used in a custom ArcToolbox and used within an open instance of ArcMap or ArcCatalog.
2. Standalone Python script - this is a script that is written with the intent to be run or executed outside of an open instance of ArcMap or ArcCatalog and may be used in a Windows scheduler program to have the script run automatically without user involvement.

This reference can be found in the ArcGIS Help under **Geoprocessing—Geoprocessing with Python—Essential Python vocabulary**.

The author makes reference of this distinction because it will impact how some Python scripts are written and error handling is developed. *A Python Primer for ArcGIS* will focus on developing standalone Python scripts, but will spend time throughout the book commenting on certain scripting methods that can be used for scripts developed for the ArcGIS Toolbox. Chapter 10 is devoted to the development of a custom ArcToolbox, developing Python script to accept user input, binding the Python script to the custom tool, and creating help documents for the custom tool. One of the major reasons for writing this book is to provide the reader an introduction to Python, use geoprocessing routines with Python, and be able to develop logical code and practice writing code to solve geoprocessing tasks and automate repetitive tasks. Creating standalone programs provides the opportunity for the developer to obtain the full experience of developing a geoprocessing task strategy, logically designing and writing the correct syntax to perform the geoprocessing routines and tasks, and to problem solve syntax and scripting logic problems that ultimatly arise in almost any geoprocessing workflow or script.

For example, a Python script can perform the following operations to create a custom data set.

1. Create a new data set
2. Add and define a specific number of attribute fields
3. Query and select records from multiple input data sources
4. Add new (features) records and calculate attribute fields to the new data set
5. Copy the features in the new data set to a different computer or server

In addition to the above, the Python script can be scheduled to run the above processes at a certain time and at a certain frequency (daily, weekly, monthly, etc) essentially automating geoprocessing tasks without user involvement.

Python and ArcGIS Versions

Certain versions of Python work with certain versions of ArcGIS and may affect how Python code is structured. Some of the methods are processed differently depending on the version of Python and ArcGIS. As mentioned above, ArcGIS 10, Python 2.6.5 and the `arcpy` module will be used in this book. Scripts developed for older versions of ArcGIS can be run within ArcGIS 10 and Python 2.6.5, provided the scripts are properly written and reference the correct version of the geoprocessing object, `arcgisscripting.create()` as opposed to the `arcpy` module used in ArcGIS 10. Some Python geoprocessing functionality has changed slightly with newer versions of ArcGIS and Python. See the following table which summarizes the geoprocessing object, ArcGIS version, and corresponding Python version. For more information consult the ArcGIS Help documentation for the respective ArcGIS version and review the Python related topics.

Geoprocessing Object	ArcGIS Version	Python Version	
`import arcpy` module No specific geoprocessing object required	10	2.6.5	Python syntax must use the required `arcpy` syntax, structures, and methods
`import arcgisscripting` module `gp=arcgisscripting.create(9.3)`	9.3*	2.5	Some Python syntax must conform to ArcGIS 9.3 Python methods (such as lists or Booleans)
`import arcgisscripting` module `gp=arcgisscripting.create()`	9.2, 9.3*	2.4, 2.5	ArcGIS 9.2 must use Python 2.4; ArcGIS 9.3 must use Python 2.5

It is recommended that newer versions (> 2.6.5) of Python are not used, since they are not supported by Esri and will likely not work.

NOTE: If the Python programmer is developing code for one version of ArcGIS, but the end user will likely use a different version of ArcGIS, the script should be tested for that particular configuration and may require a different version of Python to be installed on that system. For example, if ArcGIS 9.2 is running on a server to run nightly script processes, Python 2.4 must be installed on that system and the Python script must be able to run successfully with Python 2.4 and ArcGIS 9.2 tool syntax. For developing code, any version of Python can be used, however the Python version must match the ArcGIS software for testing and implementation and use the proper import module. Different versions of the scripts may need to be developed depending on the version of ArcGIS and Python.

*Indicates that if `arcgisscripting.create(9.3)` exists, then certain Python structures (such as lists and looping structures) will need to be written in a certain form and if the script is deployed on a different system, the machine must have ArcGIS 9.3 and Python 2.5 installed. If the `arcgisscripting.create()` is left blank, then the Python script can be written and processed with either ArcGIS 9.2 or ArcGIS 9.3, provided that the Python structures are written with the appropriate syntax for the specific version of Python.

How Python is used with ArcGIS

The author is often asked if Python be used to change the look and feel of ArcGIS or create a custom toolbar or "button" to perform a special function within ArcGIS. The answer is "no." However, a custom "tool" can be created and stored within the ArcToolbox which runs just like any other tool in ArcToolbox that is designed using Python script. Tools that use custom Python scripts often have parameters that are filled in like other tools. In addition, the new `arcpy mapping` module can provide some ability to create and generate cartographic output that previously required the use of Visual Basic for Applications (VBA) or VB .NET and a custom button and/or graphical user interface.

Python is intended to help automate geoprocessing tasks often in batch mode or through a scheduled process. Historically, Arc Macro Language or Visual Basic for Applications has served this need. Currently, Visual .NET or cross-platform C++ is used to create custom ArcGIS applications or toolsets that require a user to interact with the ArcGIS interface or mapping environment. Other resources are available to address these topics and are beyond the scope of this book. See the following website for more information http://resources.arcgis.com/content/arcgissdks/10.0/about.

Python Development Environments

A number of scripting environments exists to develop Python code. One can develop code simply by using a simple word processor or blank text file. Notepad, WordPad, Word, or other word processor application can be used.

Integrated Development Environment (IDLE)*

When a developer installs the Python application from the *python.org* site, a Python editor space, *interactive development environment* (IDE) - (IDLE), is available that allows for color coding of key words and some simple tools to assist the developer create and write code. Often when an existing script is opened with Python IDLE two windows appear:

1. IDLE - the script editor to develop code
2. Python Shell which will report back print statements and error messages when the script is run from within IDLE.

This development space also offers some basic error checking such as proper indentation and end of statements. A user can change some of the look and feel of the application, however, it is limited. The IDLE script editing environment is easy to use and does not require additional software or configuration of the editing environment. The following screen shot shows both the Python Shell and IDLE that contains simple script.

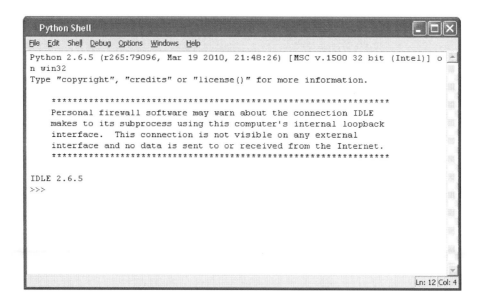

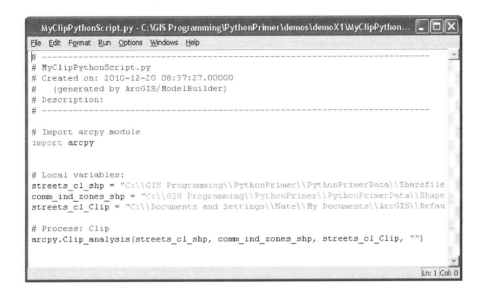

It is recommended that the Python IDLE interface be used for working through the demonstrations and exercises. The examples in this book will use the Python IDLE development environment.

*From *Learning Python*, 4th edition, pg. 58 a footnote indicates IDLE is named after Eric Idle, from Monty Python.

A number of other IDE programs are available, some of which are free, while others are commercial. Some of the other IDEs require additional installation software modules or features and are beyond the scope of this book. PythonWin, Eclipse, and Wing are such examples. The author has used PythonWin and Eclipse, but almost exclusively uses the standard IDLE interface. A short review of different IDEs can be found here.

http://wiki.python.org/moin/IntegratedDevelopmentEnvironments

Relationship to ModelBuilder

Python closely relates to Esri's ModelBuilder since a developer can use the ModelBuilder to develop workflows and generate process logic based on the ArcGIS Tools. A developer can actually spend a lot of time working in ModelBuilder until a proper set of geoprocesses, inputs, outputs, and intermediates are developed. Once a developer is satisfied with the workflow, a Python script can be exported from ModelBuilder where the developer can continue to work on and refine the scripting process. Since models only function within the ArcGIS environment, exporting a process to Python can make it more available to other non-GIS users and automated back-end processes. Chapter 2 describes how ModelBuilder can be used to develop Python script and some caveats with using ModelBuilder for code development.

Python Shell in ArcGIS

The Python Shell is also available from within ArcGIS. The user can click the Python window button to load the Python shell. Python syntax can be written and processed, including ArcGIS geoprocessing functions. Writing ArcGIS processes within this environment also provides some auto-completion of code and provides some help content for the specific geoprocessing function. Readers may find this useful to check Python syntax for developing scripts; however a Python editor will primarily be used for code development, especially standalone scripts.

The figure below shows ArcMap with the Python Shell open within ArcMap. It is opened by clicking on the Python button. Python scripts can be written and processed within this window. In addition, when ArcGIS functions are written, some code completion is available and the specific ArcGIS tool help appears on the right. Function parameters are highlighted as they are encountered when writing the Python script.

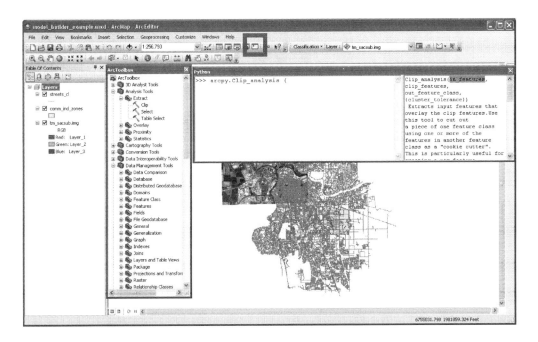

The following figure shows the Python Shell opened within ArcMap. In this example, the Clip routine is written. As the developer types in the Python syntax, the scripting window provides a list of possible geoprocessing routines. As more characters are typed in, the list becomes shorter. The developer can click on the specific geoprocessing function and the Python syntax will be completed up to this point (e.g *arcpy.Clip_analysis*).

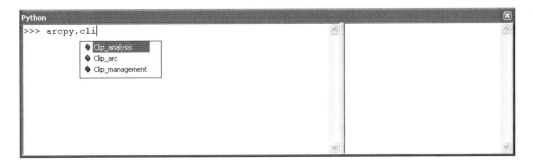

The figure below shows the Clip routine. The first parameter is highlighted on the right side indicating that the developer needs to type in a value for this parameter. The parameter may be a specific value (i.e. a specific name of a feature class). As additional parameters are filled in, the highlighted text on the right will change. When the required parameters are filled in by the developer, the ArcGIS geoprocess can be completed with an ending parenthesis. At this point Python will process the routine. Other routines and Python syntax can be written and processed. Since this code is written inside ArcMap and not in the Python IDLE environment, it will not be able to be executed outside of ArcMap.

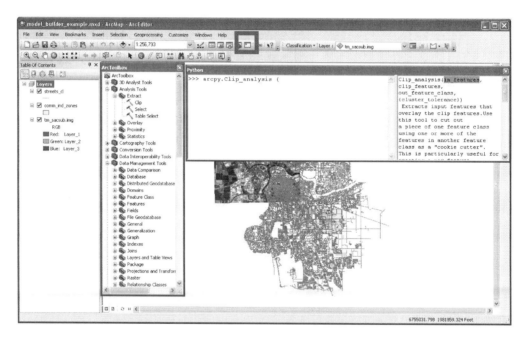

A Python Primer for ArcGIS focuses on writing code outside of the ArcGIS environment rather than developing code within ArcMap or ArcCatalog so that the reader can develop the programming and troubleshooting skills required to build functional code to perform geoprocessing tasks. Consult ArcGIS Help for more information for writing Python syntax within ArcGIS using the Python Shell.

Use of Scripts with Geoprocessing Tools

The ArcGIS geoprocessing tools found in the ArcGIS toolbox are fundamental to the development of Python scripts for ArcGIS. Essentially, developers are using the ArcGIS geoprocessing tools to create custom built automated processes that assist analysts and data managers to automate routine tasks or tasks that need to be implemented numerous times where, if performed manually, could take significant hours or days to complete. For example, an analyst could take a parcel data set from a local tax assessor group and extract a subset of data, join a number of tax and owner related tables together, and run a series of queries and computations based on landuse and the number of units found on each parcel and create a custom parcel data set that is used by a local code enforcement department that is updated each week. Performing this set of tasks could take a number of hours to do each week. A script can be developed that performs all of the above tasks to create, maintain, and update this custom dataset and it can be performed during off-peak hours. Doing so, frees up the GIS worker to focus on more analytical tasks that require direct involvement with data, analysis, and cartography.

Getting Help

A typical question of new Python developers for ArcGIS is where to get help and assistance. The Esri ArcGIS Help and support site are primary sources for gaining information and insight into Python syntax for ArcGIS and how it is used for specific tools. However, since Python is an open source application development software, some questions regarding Python may not be addressed within the ArcGIS Help or Esri support environment. Python, on the other hand, has a much broader user base than just GIS and currently, there are numerous generic clearinghouses for Python. Python for non-Esri GIS productions can be found on the web.

More advanced Python methods (e.g. dictionaries, arrays, functions, etc.) that can be used with ArcGIS geoprocessing objects will likely be researched using Internet searches for these Python methods and by consulting a Python text, the python.org website or studying other developer code found using internet searches and the help sites mentioned above.

A Python Primer for ArcGIS contains some sample scripts written by the author. Additional scripts or code snippets can be found at the author's website, www.jenningsplanet.com.

ArcGIS

Specific to Python programming, the ArcGIS ArcToolbox Help for specific geoprocesses will typically be the first point of investigation to get assistance on developing proper syntax and parameters. ArcGIS has improved the scripting help for many of its geoprocesses. In many cases, Python code developers can copy and paste code directly from the help and then modify the syntax accordingly. Much of this book will refer to ArcGIS Help documents found under **Geoprocessing**.

Secondarily, an application developer can access the online ArcGIS online resources. The online help provides the code developer essentially a world-wide GIS community. Since many people are using and developing under ArcGIS and Python, developers will often find code snippets or entire programs that provide some of the functionality they are looking for. In many cases, the developer will need to write and modify any code that is obtained off the Internet. Some of the useful resources are listed here and in the References section at the end of the book.

http://resources.arcgis.com/ - the gateway to various help, forums, and online support
http://blogs.Esri.com – blog that Esri maintains and has some useful tips and tricks and example work flows on a variety of subjects including Python development
http://forums.arcgis.com/forums/117-Python - specific web forum for Python users/developers
http://resources.Esri.com/geoprocessing/index.cfm?fa=codegallery – geoprocessing gallery of scripts and models
http://training.Esri.com/acb2000/showdetl.cfm?did=6&Product_id=971 – free overview Esri Training course on Python and ArcGIS 10

Python

As mentioned above, Python has a wide ranging user base that covers many specific disciplines. The Python Help is completely online and with a little effort a code developer can figure out how to use the Python examples. For those who like to have a hard copy text, a variety of books can be purchased on-line or at major book retailers. These can also be found at most major book retailers and online. See the References section at the end of the book for a list of resources, books, and websites related to Python and ArcGIS.

Python and ArGIS Errors

Scripting errors are inevitable when writing programming code. Learning how to understand and decipher error codes and messages will be important to trouble shooting coding and logic problems. Two groups of errors will typically be encountered when writing Python script for ArcGIS:

1. Python related syntax errors, such as typos, indentation, and Python structure
2. ArcGIS errors, which are those related to the incorrect or missing parameters or incorrect data types used in the ArcGIS geoprocessing tools, methods, and properties.

Python syntax errors will likely be identified with the Check Module routine that can be found wtihin the Python editor as well as any error message handling provided by the code developer (such as the use of print statements or the `traceback` module). ArcGIS errors will likely be identified through the use of print statements, custom error handling (using specific except: code blocks for different kinds of errors), or the `traceback` module which can identify specific ArcGIS error codes that can be referred to in the ArcGIS help. The following ArcGIS Help documents can be useful to better understand error handling with Python:

1. **Geoprocessing—Geoprocessing tool reference—Tool errors and warnings**.
2. **Geoprocessing—Geoprocessing with Python—Accessing tools—Understanding message type and severity** as well as **Error handling with Python** in the same help location.

Python Syntax Errors

Once new programmers learn some of the basic Python constructs most of the Python related errors result from mistyping, differences in capitalization with variables, indentation, and forgetting to add colons and parentheses. These will typically show up when the programmer clicks the Check Module from the Run menu in the Python Script Editor. Most of the time the line of code with the problem will be highlighted or the cursor placed at the line with the suspected problem. In some cases the problem may actually exist before the highlighted line, so the programmer will want to review the lines of code preceding the actual error. See the figure below that shows a Python syntax error after running the Check Module.

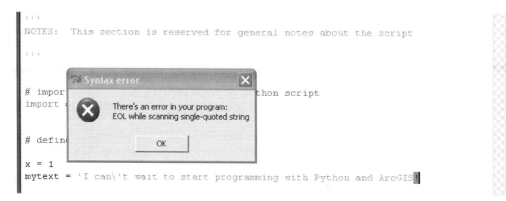

ArcGIS Error Codes

Error codes have also been a major subject of discussion in the GIS programming community and have not been very well explained. One major advancement with Esri Python scripting is the use of Esri error codes to help the developer troubleshoot and report errors back to the user (see figure below). Many of the error codes can be typed into an ArcGIS Help or ArcGIS Online search to find out more details about the specific error (e.g. ERROR 000800). See the ArcGIS Help under **Geoprocessing—Geoprocessing tool reference—Tool errors and warnings**. In many cases, the errors are related to poorly written syntax by the code developer or the incorrect information has been entered by the end user of the script or tool.

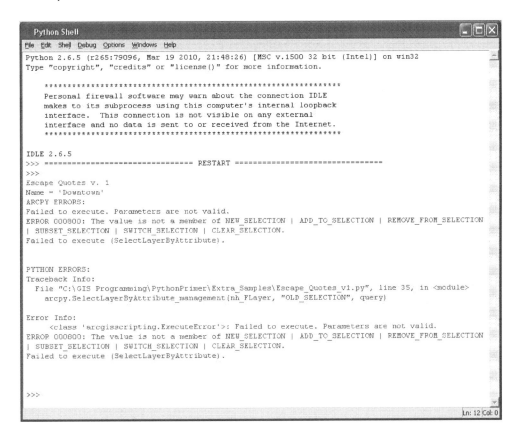

The figure below shows the Tool errors and warnings section of the ArcGIS Help document topic mentioned above. Each folder contains groups of error code numbers so the code developer can easily locate information about specific errors.

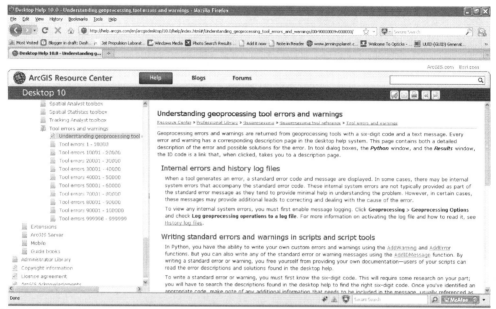

Source: ArcGIS Resource Center.
http://help.arcgis.com/en/arcgisdesktop/10.0/help/index.html#//00r90000009v000000.htm

Some Python error handling syntax can be found on several of the ArcGIS tool Help documents (usually found in an `except:` block) and be incorporated into custom code. The error syntax can often be used throughout many scripts and thus can be "recycled." The reader will find the same error handling code throughout many of the examples in this book. The error handling syntax was actually found while reviewing specific ArcToolbox Python example scripts. The exception code script can be found in the **exception.py** file located in the **Chapter01** folder of the supplemental material. See the following code snippet.

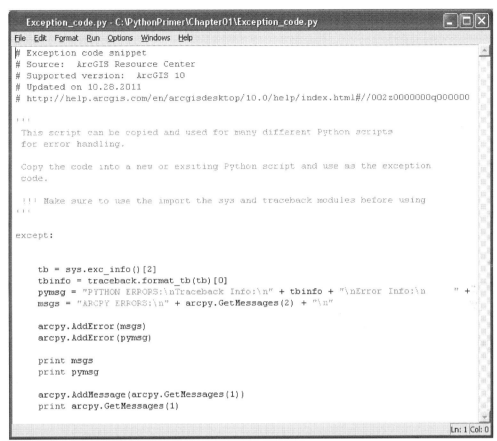

Source: ArcGIS Resource Center.
http://help.arcgis.com/en/arcgisdesktop/10.0/help/index.html#//002z0000000q000000

In addition, code developers can create specific code messages to report issues back to the user when errors are encountered. These can be reported back to the Tool progress window when the script is used as part of a custom ArcGIS tool. The author encourages the use of the *traceback* module (which is referenced in the above script and *except:* blocks) to assist with handling errors. More details about error handling are discussed throughout the book and accompanying material. Chapter 8 specifically discusses error handling.

Common Methods for Handling Errors

A number of straight forward methods can be used to troubleshoot scripting errors. Three types have already been mentioned:

1. `print` statements
2. `except:` blocks (associated with `try:` blocks)
3. `traceback` module used to capture ArcGIS Errors

The following are the most commonly used methods to help troubleshoot and will be seen throughout the book and the scripting demonstrations and exercises.

`print` Statements

Print statements can be added at any time throughout a Python script. Print statements can simply print a message out indicating that the script progressed to a certain line or it can print out a value of a variable, or a counter for a loop among others. Typically, prints statements print to the Python Shell while the code developer is working on the code. In addition, print statements can be printed to "log" files to capture progress of automated scripts. See Chapter 8 for more details.

`except:` Blocks

`except:` blocks are used in conjunction with `try:` blocks and provide an area of the code to handle errors that occur throughout the code. `except:` blocks can contain simple `print` statements or can contain more involved code that involve the use of variables and different kinds of message similar to those in the above figure.

`traceback` module

Often found within `except` blocks is the use of the `traceback` module elements such as those shown in the above figure that capture ArcGIS error messages that can be printed to a Python Shell window or to an ArcGIS progress dialog box. Exception code that reports back specific ArcGIS error messages can be very helpful when troubleshooting ArcGIS tool syntax, especially for code developers that are relatively new to the ArcToolbox geoprocessing capability.

Additional commentary regarding error handling can be found in Chapter 3. A discussion of developing custom error messages and handling can be found in Chapter 8.

Chapter 2 ModelBuilder and Python

Before jumping into Python a general overview of ModelBuilder is presented. Many current ArcGIS users are familiar with ModelBuilder and may have already been exposed to producing Python scripts from it. ModelBuilder is fairly straightforward to use and combine geoprocesses, however, developing and augmenting Python script generated from ModelBuilder can pose some additional challenges that code developers, especially those new to both Python and ArcGIS, can face that can make learning Python for ArcGIS more difficult. An overview of ModelBuilder is described and provides a general introduction to use it for developing geoprocesses. From a programming perspective the final task when using ModelBuilder is to generate a Python script where it can be further developed in a Python editor. Demo 2 outlines the specific steps to create a model and export it to a Python script. Some caveats are also mentioned at the end of the chapter with using ModelBuilder for Python script development. Some of the challenging concepts and geoprocessing (such as feature layers, table views, working with selected record sets, and looping among others) requirements will be more fully discussed in the next section.

Overview of ModelBuilder

ModelBuilder is a graphical interface often used with ArcMap where an analyst can quickly build geoprocesses using pre-existing ArcGIS Toolbox tools, other models, or script tools. ModelBuilder has been available in ArcGIS for a number of years and many GIS professionals are familiar with and use ModelBuilder to generate multi-step geoprocesses for tasks such as site suitability, site assessment, the movement of materials over a landscape, and changes over time, among others. Before Python was available, ModelBuilder was the primary method for generating multi-step geoprocesses. Since *A Python Primer for ArcGIS* focuses on developing Python script and code development, this chapter describes some basic ModelBuilder operations with the primary goal of generating a Python script that can then be modified within a Python script editor.

Esri has continued to make improvements and add functionality to ModelBuilder such as looping, conditional statements, and being able to embed other models and scripts. Many geoprocesses can be developed, built, and run completely within ModelBuilder without the need to write any Python script. If the only use of a geoprocess is to use it within ArcGIS or embedded within an ArcGIS Server environment, then ModelBuilder will meet this need. If, on the other hand, a GIS analyst wants to be able to use and process data from different environments, reset workspaces throughout a process, and auto run geoprocesses without any user interaction or dependency of ArcGIS being open or ArcGIS Server using the model, then developing standalone Python scripts will meet these needs and may be easier to develop and make available for other non-GIS users to integrate their own back-end or scripting routines.

For a thorough discussion of ModelBuilder, readers are encouraged to refer to the ArcGIS Help documents under **Geoprocessing—Geoprocessing with ModelBuilder**. *A Python Primer for ArcGIS* will refer to some of the basic concepts as well as being able to export a model to Python script. Later, Chapter 10 will describe how to integrate a Python script into ArcToolbox, similar to how a ModelBuilder model can be run within a custom ArcGIS Toolbox.

ModelBuilder Python Script Caveats

Developing geoprocesses in ModelBuilder with the intent of generating Python script can pose some frustrating challenges to the code developer if the code developer is not aware of some of the nuances of how ArcGIS references and uses data, workspaces, and data paths. Using data from the Table of Contents versus accessing data directly from disk can result in different Python syntax. When code developers try to run a standalone script without specific paths set to data, the script will likely result in error. This can be a large issue for more complex models that are then exported to Python script and is the primary reason why the author cautions the reader about using ModelBuilder for developing Python scripts.

The figures below show two different scripts pointing to the same data on disk with the only difference being that the first script uses data that has been browsed to on disk when setting the Clip tool parameters, whereas, the second script references the data in the Table of Contents.

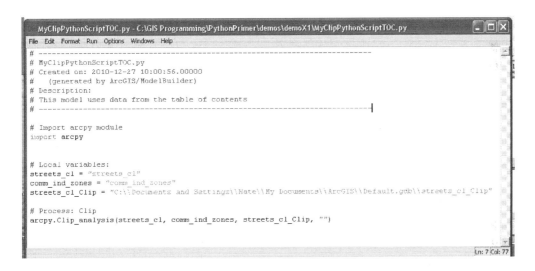

The first script will execute properly when run as a standalone script. The second script will result in error because Python does not know the location on disk where "*streets_cl*" and "*comm_ind_zones*" are or recognize the data format. These values actually represent "feature layers" that point to the actual "feature class" (i.e. a shapefile or feature class in a geodatabase) on disk. Feature layers are virtual representations of actual data on disk. This is another issue when developing Python scripts from ModelBuilder.

The second script generated out of ModelBuilder does not include the "Make Feature Layer" function which is required to create the feature layer from the feature class. By default, when data is loaded into the Table of Contents,

ArcMap automatically creates the feature layer and thus can be used in a model or other geoprocess when ArcMap or ArcCatalog are opened. When creating standalone scripts other pre-requisite steps may be required for the script to function properly (e.g. defining data paths, workspaces, feature layers, and table views among others). The programmer needs to be aware of the limitations and restrictions of ModelBuilder when using it to develop Python scripts, especially those intended for standalone implementation.

It has been the experience of the author that using ModelBuilder is a good tool to assist with developing geoprocessing workflow and the general set of steps as well as some of the Python scripting logic for standalone scripts. However, for making on-going changes and modifications to the Python script, it has been more beneficial and efficient to continue to work within a Python editor versus attempting to modify the ModelBuilder structures to then "regenerate" a new Python script. One of the objectives of *A Python Primer for ArcGIS* is to assist the developer to generate and modify Python script that can be used more broadly with GIS and data management tasks (i.e. standalone scripts). The major focus of this text is to illustrate how to use the major ArcGIS geoprocessing constructs with Python syntax to generate such scripts and build the developers GIS and Python scripting knowledge to accomplish the most commonly used GIS geoprocessing tasks.

A script can be generated from ModelBuilder using these steps:

1. Create a New Model
2. Add geoprocessing tools from ArcToolbox
3. Fill in the parameters for each tool
4. Save the model
5. Test the model
6. Refine the model as required
7. Export model to Python script
8. Make additional edits in a Python script editor

Demos Chapter 2

The demos for Chapter 2 provide a basic overview of using ModelBuilder to with the goal of exporting the model to a Python script. **Demo 2a** will focus on the primary methods to add ArcGIS Tools and generate the Python script. **Demo 2b** focuses on create a Clip model using data from disk as well as from the Table of Contents. Python scripts will be generated from each model and reviewed using Python IDLE.

Demo 2a: Using ModelBuilder to Create a Python Script

This demonstration provides a general overview of using ModelBuilder to generate a Python script. **Demo 2a** also outlines the common steps required to start ModelBuilder and associate it with a unique ArcGIS Toolbox so that it can easily be found for future use. Custom models and Tools can be saved and provided to others to use.

A couple of different methods exist to create a custom model.

Method 1: Start ModelBuilder from the ArcMap or ArcCatalog interface.

Start ModelBuilder

1. Click the ModelBuilder Button within ArcMap (or ArcCatalog)

An empty model appears.

Method 2: Create a Custom toolbox and add a New Model.

Create a Custom ArcToolbox

A custom ArcToolbox is an easy way to keep custom models or scripts organized. Custom toolboxes can also be saved and provided to other users.

1. To create a custom toolbox, right click on the ArcToolbox and choose Add Toolbox.

A user can choose to create a toolbox in the default location for the map or choose a different location for the new toolbox. To have the toolbox available for all users of the map document, the new toolbox must be placed where the other ArcToolboxes are located (i.e. **C:\Program Files\ArcGIS\Desktop10.0\ArcToolbox\Toolboxes**).

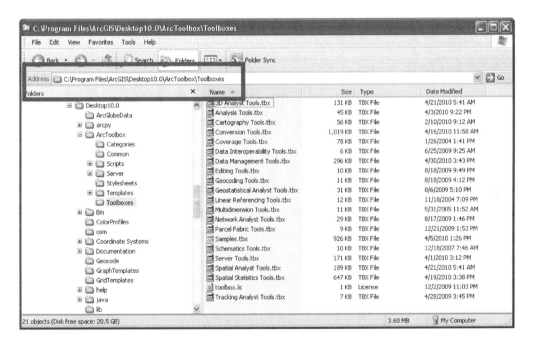

After the user chooses AddToolbox, the following dialog box appears.

2. Click on the New ArcToolbox button shown above.

3. Provide a new name for the new toolbox (e.g **MyCustomToolbox**).

4. Click Open to add the new toolbox to the ArcToolbox.

5. To add a new model to the toolbox, right click on the new toolbox created above and choose **New—Model**.

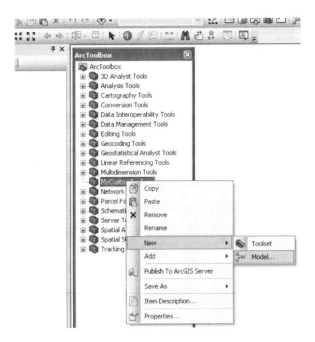

With either method to create a new model, the following screen is shown.

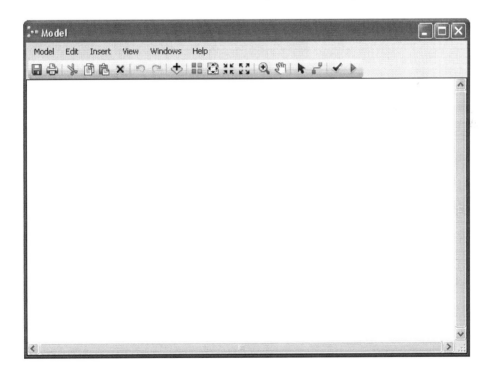

An empty ModelBuilder window appears. Various tools are shown at the top of the screen to add data, zoom in/out, pan the model, make connections between model objects, validate, and run the model. Refer to the ArcGIS Help for a more detailed discussion of each button and option (ArcGIS Help documents under **Geoprocessing—Geoprocessing with ModelBuilder**).

Adding a Geoprocessing Tool to a Model

Typically, the person designing the model will select geoprocessing tools from ArcToolbox and drag them into the ModelBuilder window. In the example below, the user clicked the Clip tool and dragged it into the ModelBuilder window.

Add a Geoprocesing Tool to ModelBuilder

Select a geoprocessing tool to the model. The user can drag and drop a tool from the ArcToolbox to the model. The Clip routine is shown below.

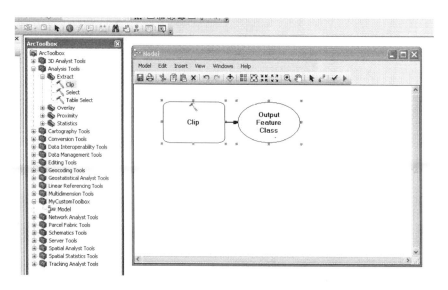

The Clip routine is shown in a "non-ready" state indicating that it cannot be run without some additional information. In this case, the model needs some input data, tool parameters set, and an output.

Setting Parameters for a Geoprocessing Tool

The user can double click on the Clip routine to bring up the tool parameter dialog box. The specific parameters can be filled in. The user can choose data from the Chapter 2 folder to fill in the parameters.

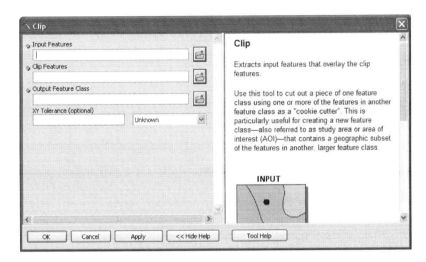

Once the parameters are set and clicking OK on the tool, the appearance of the model will change. Use the pan and zoom in/out tools as needed to move or re-center the model.

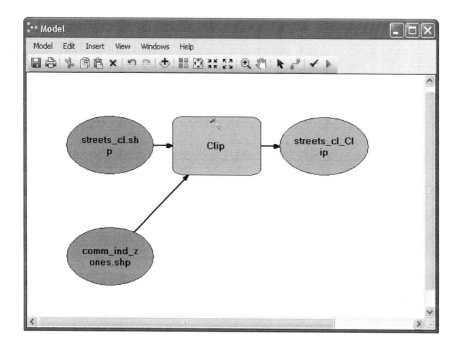

In addition, select on each tool to expand each input or output to make it more readable.

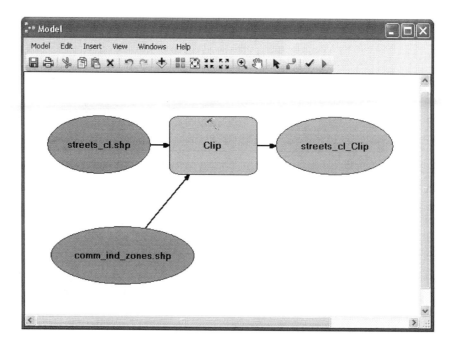

Notice that the model has inputs and an output and the model objects are colored. This indicates that a model is in a "ready" state and can be run.

Rename and Save the Model

Before saving the model, the default model name needs to be changed. Click on **Model—Model Properties**. Rename the model. Click OK and then click Save from the Model menu. The model in the custom ArcToolbox will show the new name. NOTE: The name must not include any spaces, underscores, or special characters. The label can have spaces, underscores, and special characters. The label appears in the ArcToolbox as the model name. A description can be added as well which will show up in the tool help.

Testing the Model

Before exporting to a Python script, the model should be run to determine if it will provide the desired results. Since ModelBuilder is often run within an ArcMap session, the user can check to make sure the proper input feature classes and parameters are set up as well as check the output to determine if the results of the model are correct. This can help reduce issues with data sources and output before working on code refinements.

Generate the Python Script

Once a model has been tested and reviewed, the model can be exported to a Python script. Click on the Model menu and then choose **Export—To Python Script.** Choose a location and file name for the Python script. The extension for the script will be (*.py*).

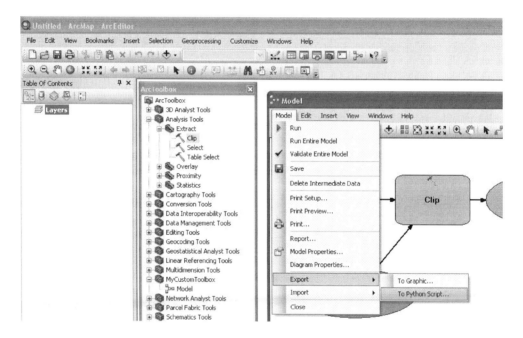

Modify Python Script

The script created above can be opened in a Python script editor and modifications can be made. To open the script, first start a Python editor and then browse for the Python file created above. Alternatively, the developer can browse for the script and right click on the file name to open with a Python editor.

Either method for opening a Python script will automatically provide the Python Shell window and a separate script window showing the Python code.

ModelBuilder Generated Python Script Commentary

Notice above that all of the information to run the geoprocess is provided. An area of comments is provided that includes the name of the Python script, the date the script was created, and a description if the developer added a description to the model. The `arcpy` module line is added as well as a section for default variable names to the data paths for the geoprocessing tool. Finally the actual geoprocessing line is added which uses the variable names.

The basic structure of a Python script is provided. The code developer can now make modifications to the code as desired. For example, the default variable names may need to be changed, relative paths set up for accessing data, adding looping structures, try except blocks, and adding custom print statements and error handling code may be desired.

Demo 2b: Build the Clip Model Two Ways and Export to Python Script

This demonstration follows the process described above to generate the Python Script for the ArcGIS Clip routine. Two different Clip models will be created.

1. Clip model setting parameters with data on disk
2. Clip model setting parameters with data from the Table of Contents

The data used in this demonstration are:

   a. Sacramento_Streets
   b. Downtown

Open **Demo2.mxd** from **\PythonPrimer\Chapter02** folder. Load the Sacramento Streets and Downtown neighborhood boundary into the Table of Contents if they do not appear. All of the data for this demo are in shapefile format and can be loaded from the **\PythonPrimer\Chapter02\Data** folder.

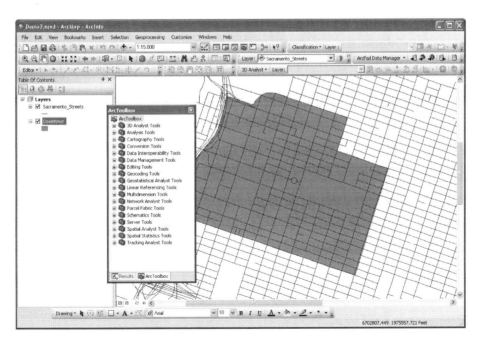

A Python Primer for ArcGIS®

1. Create a new custom ArcToolbox by right clicking on the **ArcToolbox—Add Toolbox**. Name it **MyCustomToolbox**. The toolbox can be placed in any location for the demo or in the **Chapter02** folder.

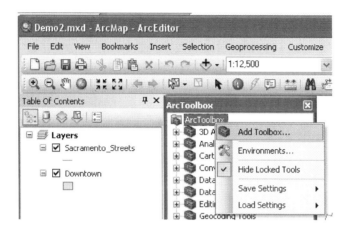

2. Open a New Model from within the new custom toolbox.

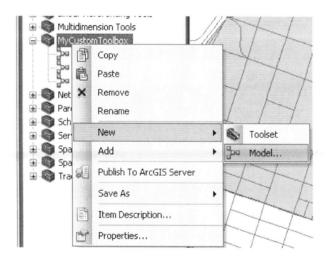

3. Add the **Clip** tool from the **Analysis Tools—Extract toolset** to the new model.

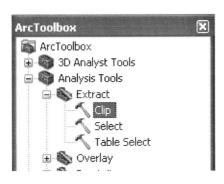 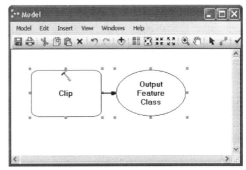

4. Double click the Clip tool and fill in the tool parameters.

The *Input Features* will be the **Sacramento_Streets.shp** file
The *Clip Features* will be the **Downtown.shp** file.
Provide an output shapefile name within the **\PythonPrimer\Chapter02\Data** folder.

For this step use the full data path by clicking on the file browser (shown below) in the tool. In the next model, you will use data from the Table of Contents.

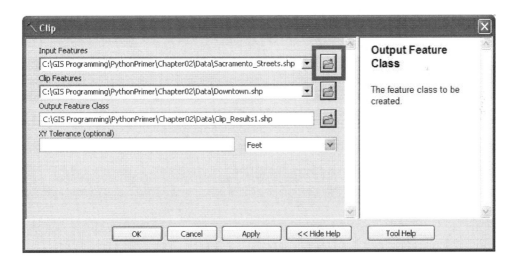

5. Change the name of the model to **ClipData1** (Name) and **Clip_Data_1** (Label), in the Model Properties.

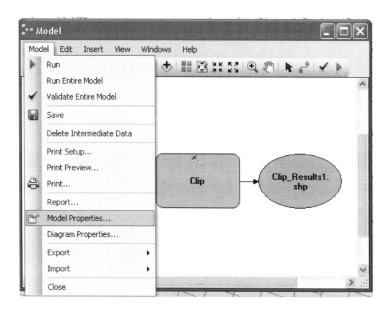

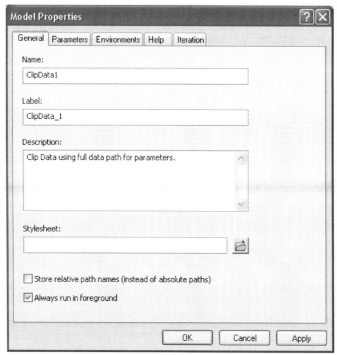

6. Save the model.

7. Run the model to make sure it runs.

8. Export the model to Python script. Name the script **Clip_Data_1.py** and put in the **\PythonPrimer\Chapter02\Data** folder.

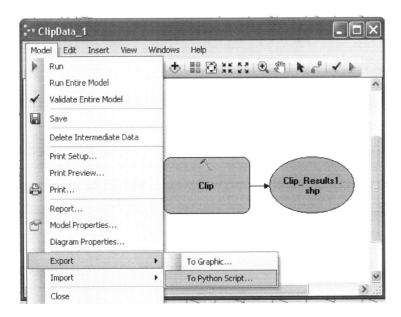

9. Open and view the script in Python IDLE. Notice the variable definitions and how they are used in the Clip routine.

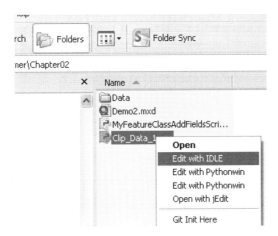

```
# ----------------------------------------------------------------
# Clip_Data_1.py
# Created on: 2011-01-03 14:49:08.00000
#    (generated by ArcGIS/ModelBuilder)
# Description:
# Clip Data using full data path for parameters.
# ----------------------------------------------------------------

# Import arcpy module
import arcpy


# Local variables:
Sacramento_Streets_shp = "C:\\PythonPrimer\\Chapter02\\Data\\Sacramento_Streets.shp"
Downtown_shp = "C:\\PythonPrimer\\Chapter02\\Data\\Downtown.shp"
Clip_Results1_shp = "C:\\PythonPrimer\\Chapter02\\Data\\Clip_Results1.shp"

# Process: Clip
arcpy.Clip_analysis(Sacramento_Streets_shp, Downtown_shp, Clip_Results1_shp, "")
```

10. Create a new model using the same methods above.

Rename the model **ClipData2** the *Name* property and **Clip_Data_2** for the *Label* property.

Instead of using the full data paths to the data, use the data in the Table of Contents. The output file can be set to some shapefile name in the **\PythonPrimer\Chapter02\Data** folder.

Save the model. Run it if desired.

Export the model to a Python script called **Clip_Data_2.py**.

Open and view the script. Compare the two scripts for similarities and differences. **Clip_Data_2.py** should not have a data path shown with the variable definitions.

```
# ----------------------------------------------------------------------
# Clip_Data_2.py
# Created on: 2011-01-03 16:05:46.00000
#    (generated by ArcGIS/ModelBuilder)
# Description:
# This model uses data from the table of contents
# ----------------------------------------------------------------------

# Import arcpy module
import arcpy


# Local variables:
Sacramento_Streets = "Sacramento_Streets"
Downtown = "Downtown"
Clip_Results2_shp = "C:\\PythonPrimer\\Chapter02\\Data\\Clip_Results2.shp"



# Process: Clip
arcpy.Clip_analysis(Sacramento_Streets, Downtown, Clip_Results2_shp, "")
```

Exercise 2: Create a New Feature Class and Add Fields

In this exercise, the user will create a new model that creates a new feature class and add some fields to the new feature class using existing ArcGIS Tools. Use the **Create Feature Class** and **Add Field ArcGIS** tools. Research the ArcGIS Help as necessary to compile a model. Test the model and export to Python script. Test to see if the script runs from Python IDLE. Note any problems, issues, or successes. Answer the questions below.

Requirements

1. In the custom toolbox created above, create a new model for Exercise 2.

2. Give the model a unique name, e.g. **CreateFeature_AddFields**

3. Add the ArcGIS tools mentioned above to the model with the following conditions.

    a. Create a feature class type of your choice (point, line, or polygon)

    b. Add several attribute fields to the table. Use a mix of data types (text or numbers). For text fields, provide a length. For number fields (short or long integer, provide a precision

    c. Make sure to connect the Add Field tool(s) to the Create Featureclass Tool or other Add Field tools in the model.

4. Use data path locations from the tool's browser for parameters. Do not use data that is already present in the ArcMap Table of Contents.

Chapter 2: Questions

1a. What is the full toolbox organization (i.e. Toolbox—Toolset—Tool) for the Create Feature Class?

1b. What are the required parameters for this tool?

2a. What is the full toolbox organization for the Add Field tool?

2b. What are the required parameters for this tool?

3. How many times do you need to use the Add Field tool?

4. What is required to set the feature class location?

5. What kind of feature class location did you use for the Create Featureclass tool?

6. In the Add Field tool help, can you set the precision and scale using a feature class in a file or personal geodatabase?

7. After creating the model, did the model run? Yes/No. If not, describe some of the methods used to attempt to fix the model? Did your modifications fix the model?

Export the model to a Python script (even if you were not able to fix it).

8. What modules were imported for the script?

9. How many "Local Variables' were created for this script? Do any of them refer to the same value? If so, which one(s)? Do you think any of the variables can be eliminated? Why/Why not?

Chapter 3   Python Constructs

Overview

A number of basic Python programming fundamentals need review before writing Python script. These fundamental will be used extensively when developing code and will likely cause significant frustration if they are not strictly adhered to. This section covers the basic and most commonly used Python structures when programming Python scripts for ArcGIS. Consult a Python book (such as *Learning Python*) or visit the *python.org* site for more details and for a more comprehensive discussion of these structures as well as others.

Using Python IDLE for Code Development

As mentioned above the Python *interactive development environment* (IDLE) is provided as part of the Python install and will be used throughout this book for demonstrating code. It is recommended that this interface be used when working through the programs within the book as well as with the reader's own code.

Starting Python IDLE

When a code developer clicks on **Program Files – IDLE (Python GUI)** from Windows, the Python Shell appears.

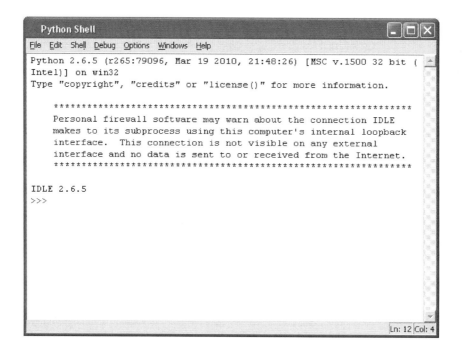

The code developer can go to **File—Open** and then browse for an existing Python script. For a new script, choose **File—New Window**. In either case a new Python scripting window appears. This is where Python script will be written and edited.

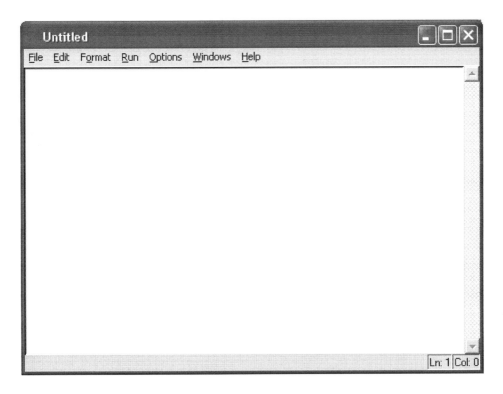

Alternatively, if a Python script file is found in the Windows Explorer, the user can right click and choose "Edit with IDLE" which will bring up a new Python Shell and load the script into a Python script editor. See below.

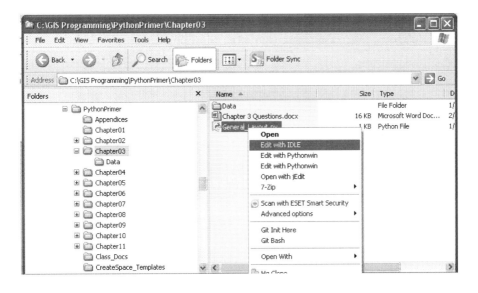

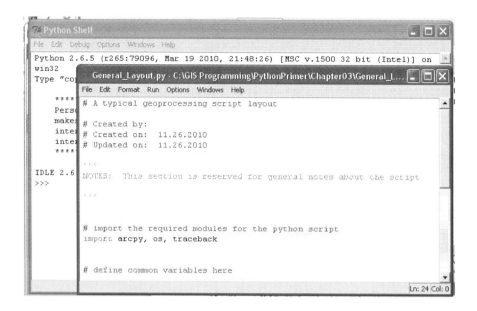

At this point a code developer can begin to write or edit code. Before getting started, read through the rest of this chapter to get an overview of the common elements of Python programming and its organization. At the end of the chapter, the reader will have an opportunity to write a "First Script" that demonstrates some of the concepts described in this chapter.

Using the Python Shell for Code Testing

The code developer may find it useful to type in specific lines of Python script without writing, saving, and troubleshooting code. Python script can be written interactively in the Python Shell window and see results. For example, the figure below shows some variables and an "if" statement typed into the Python Shell showing the results immediately after typing the lines. Notice that error messages will pop up when syntax is written incorrectly.

```
Python Shell
File  Edit  Shell  Debug  Options  Windows  Help
IDLE 2.6.5
>>> x = 1
>>> print str(x)
1
>>> astring = "This is a string"
>>> print astring
This is a string
>>> if x == 1
SyntaxError: invalid syntax
>>> if x == 1:
        print atring


Traceback (most recent call last):
  File "<pyshell#7>", line 2, in <module>
    print atring
NameError: name 'atring' is not defined
>>> if x == 1:
        print astring


This is a string
>>>
```

The Python Shell can also be accessed and used within ArcGIS as described in the previous chapters.

Syntax

Python is a fairly flexible scripting language, provided the developer follows a few coding rules. Any comprehensive Python book as well as the *python.org* site above contains hundreds of pages on all of the various constructs Python has to offer, however, most of the commonly used coding structures are needed to write most ArcGIS Python scripts.

Case Sensitivity

Python code developers need to pay attention to the capitalization and naming of variables, values, and other programming constructs. Typing the same name using a mix of capitalization will be interpreted by Python as distinct names or values. For example, the two names below that show different capitalization need to refer to different values (in this case, two distinct feature class shapefiles). Using the same name with different capitalization can lead to difficulty in troubleshooting code and may produce unintended consequences.

```
featureclass = "aFeatureClass.shp"
FeatureClass = "aDifferentFeatureClass.shp"
```

Naming Conventions

A code developer should adopt a naming convention for a program. This will help code become more readable to the developer and the end user and will help the code developer find and debug problems.

Common examples can use a lower case abbreviation for a data type and then a name starting with an upper case letter. For example,

```
strFieldname  =   'parcel'
```

may represent a variable name that is set to the word "parcel" (i.e. a string of characters).

The developer can also simply use:

```
fieldname = 'parcel'
```

where the variable name still represents the word "parcel." The programmer should maintain a standard practice of naming objects. The name should be meaningful, represent the object, and possibly the type of data (e.g. strings of characters, numbers, feature classes, images, tables, fields, etc.).

Indentation

Indentation is a key requirement when writing Python script. Indentation tells Python that a set of code will be processed as a contiguous block. Indentation is used when implementing looping structures such as `while` and `for` loops), `try: except:` blocks, and conditional statements such as `if` or `else` statements. The figure below shows such an example. Notice that the both the `for` loop and the `if` statement show indentation. The indented lines of code are considered a "block of code" and processed line by line.

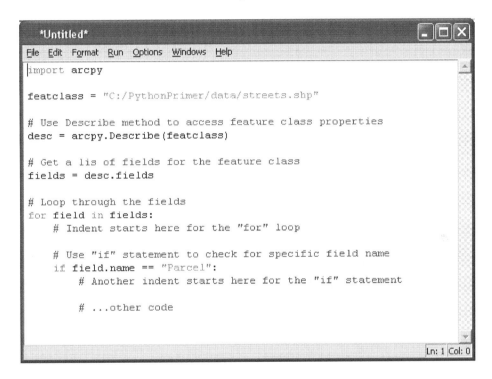

Depending on the kind of Python script editor the programmer uses, the Python IDLE script editing tool does a good job of properly indenting code. The developer can also use the Check Module before running the code that may identify improperly indented lines. Consistent indentation is important. Maintaining consistent indentation is important when developing code because Python interprets the indentation as a contiguous block of code. A "tab" or a specific number of spaces (e.g. 4 spaces) indicates an "indent." The default indentation is a "single tab" or four spaces and can be modified in the **Python—Options—Configure IDLE** properties for the Python IDLE script editor. For example, if indentation consists of four (4) spaces or a single tab, then this format should be used throughout the script. Some Python editors provide a setting to modify any indentation defaults. See the specific Python editor help regarding indentation settings.

Comments

Comments are used to provide in-line documentation and commentary about the code. Comments can take the form in two different ways. The figure below shows both methods being used in a Python script.

```
# Exercise 1

# Created by: <author's name>
# Created on: 11.26.2010
# Updated on: 01.02.2011

'''
NOTES:  This section is reserved for general notes about the script
Multiple lines can be written here or the triple quotes can be used
to comment a block of code.
'''

# import the required modules for the Python script
```

Comments are helpful because they often provide useful notes about the program, can be used to create an outline for the script, and provide some additional information about program variables and geoprocesses.

Creating and Using Variables

Python often requires defined terms, called variables that the programmer uses to write concise and flexible code. A variable is essentially a name that is assigned a specific value that can be used in different parts of the code. For example, a variable can represent different feature classes that are used as an input to a "clip" geoprocessing routine. The code below illustrates the use of variables to define feature class names and the buffer distance for the ArcGIS Clip routine.

```
in_fc = 'c:\\temp\\aFeatureClass.shp'
clip_fc = 'c:\\temp\\aPolygon_FC.shp'
out_fc = 'c:\\temp\\anOutFC.shp'
fuf_dist = '200 FEET'

arcpy.Clip_analysis(in_fc, clip_fc, out_fc, buf_dist)
```

The defined variables in_fc, clip_fc, out_fc, and buf_dist can be used in any place in a script rather than write the full path an shapefile names and the specific buffer distance each time the shapefiles and distance are required in the Python script and ArcGIS geoprocessing functions. Variables can be assigned to almost any value or reference. Variables can be assigned to numbers, character strings, dates, etc. Variables can also be assigned to data paths, work spaces, tables, feature classes, data bases, feature data sets, feature layers, file names, etc. In many cases, the variables are assigned to numbers or character strings (e.g. a buffer distance, a specific value such as an index number or counter, a table or feature class name, a work space, a query, or a file name). The figure below shows some of these examples.

```
import arcpy
from arcpy import env    # environment module

# Some examples of common variables used with ArcGIS and Python

env.workspace = "c:\\temp\\"     # an ArcGIS workspace

datapath = "c:\\temp\\"                       # a directory (folder)
inshapefile = datapath + "parcels.shp"        # a shapefile in the c:\temp director
pGDB = "c:\\temp\\pGDB.mdb"                   # a personal geodatabase
infeatureclass = pGDB + "\\" + "streets"      # a pGDB feature class


# String examples

aName = 'ESRI'                   # a text string variable assigned the value
aShapeFileName = "parcels.shp"   # a text string representing a shapefile nam
query = "[APN] = '0123456789000'"  # a text string to define a query (where cla
fLayer = "parcel_layer"          # a text string that can be used to "make" a feature l


# Number examples

x = 0    # a variable x assigned the value 0 (e.g. to initialize a loop counter)
aFloatNumber = 4.5  # a decimal number

y = x + aFloatNumber   # a variable y that equals the sum of two numbers
                       # (in this case 4.5)

buffDist = 100    # this variable can be used in a buffer routine

arcpy.Buffer_analysis (<input_fc>, <output_fc>, buffDist + " Feet")
```

String and Number Variables

Some of the most common uses of variables are to define strings (i.e. a series of alpha numeric characters) for names, queries, feature layer names, table view names, join table names, specific field names, etc. Strings are enclosed in single or double quotes. Numbers may also need to be assigned to variables so that they can be used and changed as needed. Some common uses of number variables include (number calculations, counters, values in ArcGIS geoprocessing tool parameters such as distance, area, unit measures, etc). Numbers do not have quotes around the value.*

*NOTE: Some numbers may be "cast" as strings so that they can be used in query strings, print statements, and ArcGIS geoprocessing parameters that require strings that have both numbers and text characters. For example, to cast a number as a string, the following syntax can be used `str(<a number>)`. Consult other Python texts and the Python website for more details on casting values.

The reader will find examples throughout the demo scripts and exercises that use the `str()` routine to cast numbers as strings. Both strings and numbers make up the majority of variable definitions used in ArcGIS parameters.

Lists

A Python list is a special structure that can store and index a number of related elements (e.g. a list of names, a list of numbers, a list of fields in a feature class, feature classes in a folder or geodatabase, an image in a list of images (rasters), etc). A set of list data methods are available in ArcGIS. See Chapter 7 in this book and the ArcGIS Web Help under **Listing Data** for more information on creating and using lists.

In general Python creates lists in this manner:

aList = [1, 2, 3, 4, 5]

The values in the [ ] represent specific elements in the list. The values in a list are accessed by obtaining the index (i.e. the location of the value in the list. Python lists indexes begin with the value of zero (0), so the "first" element (`aList[0]`) in the list is 1.

To access an element, the following can be written:

```
aList[1]      # the value in the list at second position in the list is 2
print aList[3]   # will result in the number 4 printed to the screen
```

The figure below shows how a simple Python list can be created and used.

```
Python Shell
File Edit Shell Debug Options Windows Help

Python 2.6.5 (r265:79096, Mar 19 2010, 21:48:26) [MSC v.1500 32 bit (In
tel)] on win32
Type "copyright", "credits" or "license()" for more information.

    ****************************************************************
    Personal firewall software may warn about the connection IDLE
    makes to its subprocess using this computer's internal loopback
    interface.  This connection is not visible on any external
    interface and no data is sent to or received from the Internet.
    ****************************************************************
IDLE 2.6.5
>>> aList = [1,2,3,4,5]
>>> print aList[0]
1
>>> print aList[3]
4
>>>
```

As mentioned above, a list can contain numbers, names, fields, etc. A standard Python list can be used in this case; however, ArcGIS provides some special list methods that can be used directly. For example, the following "creates" a list of fields from a feature class.

```
fieldlist = arcpy.ListFields("parcels.shp")
```

The following looping structure can then be used to operate on each element in the list.

```
for field in fieldlist:
    print field.name
```

A more thorough discussion of lists for ArcGIS and Python is discussed in Chapter 7. Code developers may find the standard Python list structure useful in some scripts. The author's website www.jenningsplanet.com has some examples of Python lists being used in ArcGIS Python scripts.

Conditional Statements and Loops

Another common set of Python structures is the use of conditional statements and looping structures. This section introduces the concepts; other chapters will demonstrate the use of these structures. Conditional statements provide a means for decision making to take place within a script based on one or more conditions. Looping structures allow a block of code to "iterate" multiple times, often with a set of conditions so that the block is not "unbounded" or is able to "stop" after a limited number of iterations. If loops are not "bounded," the code can run indefinitely, which is not desired. Notice also that all conditional and loop statements use indentation as part of the coding syntax.

Three kinds of conditional and looping structures are used in Python:

1. **If Statements** – used to "test" a condition or set of conditions. Optional statements include the use of `elif` and `else` statements.

    The general form of an `if` statement is:

    ```
    if <condition is true>:
        # this block of script is processed
        # can be multiple lines

    elif <a different set of conditions are true>:
        # this block of script is processed
        # can be multiple lines

    else:
        # this block of script is processed (if all other if
    and elif conditions fail)
        # can be multiple lines
    ```

2. **While loops** – used to iterate a block of code "while" a conditional statement is true.

    The general form of a `while` statement is:

    ```
    while <condition is true>:
        # process a block of code
        # can be multiple lines
    ```

3. **For loops** – used to iterate a block of code for each element in a sequence or group of elements.

```
For <item> in <sequence>:
    # process a block of code
    # can be multiple lines
```

The conditional and looping structures are introduced here to show the general syntax. Other chapters will discuss and illustrate these structures in more detail.

import Modules

Python requires specific modules to be imported (i.e. provided to the Python processes at the time of running a script) so that different operations can take place (such as running ArcGIS functions, string processing, math, or operating system functions). The first "real" lines of code shows the modules that are "imported" into the Python processing environment so they can be used throughout the script as required. The developer must import those modules using the "import" command. ArcGIS and the proper version of Python must be installed on the system where the ArcGIS Python script is executed. Data sources (feature classes, tables, geodatabases (except SDE) can exist on computers, servers, and networks, that do not have ArcGIS installed on them.)

```
import arcpy, os, sys, traceback
```

The above line imports the following Python modules that will be used in a Python script.

arcpy – provides access to all of the ArcGIS geoprocessing functions including the mapping (`mapping`), Geostatistical (`ga`), and Spatial Analyst (`sa`) modules. Pre-ArcGIS 10 uses the `arcgisscripting` module. See Chapter 1 for specific notes using the `arcgisscripting` module, Python versions, and ArcGIS versions. The `arcpy` module is required to perform any ArcGIS geoprocessing.

os – provides access to operating system functionality such as file and directory path operations. The `os` module is optional in many cases, but required for some file, folder, and data source management operations that use standard Python syntax. Consult a Python text or website for more details.

sys – is used to access by Python system functions and are often found in defining variables for user input such as:

```
myshapefile = sys.argv[1]   # variable accepts user input for
                            # the first real argument or
                            # parameter from a custom
                            # ArcGIS tool that uses
                            # a Python script
```

The `sys.argv[]` structure will be discussed in more detail in Chapter 10 which reviews how create and implement a custom ArcTool that uses a Python script.

traceback – is used to for error handling; this module is not required, but it is often useful for handling errors and the same blocks of code can be reused in many Python scripts

For most ArcGIS Python scripts developed in this book only the `arcpy` module and `traceback` will be required (because of error handling). Consult a Python text or the Python website for a full account of Python modules.

try: and except: Blocks

`try:` and `except:` blocks are commonly used to group lines of code so that error handling can occur. Scripts can be written without the use of `try:` and `except:`; however, troubleshooting code and process errors can be more difficult to remedy. `try:` and `except:` must be used together. Proper indentation is required so that Python knows to process the lines of code as a block. For this book, the most common implementations of `try:` and `except:` will be used. Consult a Python text or the Python website for a more in depth discussion of `try:` and `except:`.

Special Considerations for Strings in Python

The reader will see in this book as well as ArcGIS and Python help sites and user forums a variety of Python syntax with strings using backslashes, forward slashes, and single and double quotes, and "escape" characters. A brief discussion of how each of these is used is provided below. Additional commentary can be found throughout the rest of the book, especially in Chapter 5 which discusses queries and selecting data. In addition, the reader can consult a Python text or the Python website where these topics are covered in more detail.

Backslash, Forward Slash, and Raw Strings

Python typically translates characters literally unless specifically translated into different forms (e.g. a number value is translated into a character, `str(count)`, where `count` is a number). As a result, character strings can be challenging to work with, especially when defining workspaces, directory paths, query strings, or various parameters for ArcGIS tool parameters.

To help Python make sense of some of these character strings, a set of "escape" characters have been devised that can be used by the code developer that allows Python to appropriately translate the strings. See http://docs.python.org/reference/lexical_analysis.html for a complete list of these characters.

The single backslash ("\") acts as an escape character that can be used in a number of instances. For example, if a workspace path needs to be coded, the following Python syntax will work:

```
arcpy.env.workspace = "C:\\temp\\aPersonalGDB.mdb"
```

Notice the first ("\") "escapes" the second ("\"). This allows Python on a Windows operating system to interpret the path to the file correctly:

```
C:\temp\aPersonalPGD.mdb
```

Alternatively, the same string can be written as follows using "raw string suppression" syntax, where the letter "r" is used in front of the quoted character string so that Python interprets the single backslashes correctly on a Windows operating system.

`arcpy.env.workspace = r"C:\temp\aPersonalGDB.mdb"`

A third option for defining directory or workspace paths is to use the forward slash ("/"). The same path above can also be written like this:

`arcpy.env.workspace = "C:/temp/aPersonalGDB.mdb"`

Without the "r", the double backslash or the use of the forward slash, Python would interpret the file path as `c:<tab>emp\<bell character>PersonalGDB.mdb`. The "\t" translates to a tab and "\a" translates to the "bell" character.

As previously recommended, code developers should consistently use one of the above methods when using strings in directory paths, workspaces, queries, and ArcGIS function parameters.

Single and Double Quotes

Single and double quotes are used to enclose character strings and operate the same; however, if single quotes are used in string names, it is recommended that the single quote be "escaped" as described above. In some cases the quotes will require "escaping." This issue is often encountered when developing query strings for the "where clause" parameter for several of the ArcGIS functions such as `SelectByAttribute` and `SearchCursor`. The figure below shows a script using double and single quotes as well as the escape character to form examples of a query string that might be used in a `SelectByAttribute` function. The script can be found in the **\PythonPrimer\Extra_Samples\Escape_Quotes1.py** file.

```
# variable with a text string using double quotes
txtString = "Name"

print txtString
print "\nThe above is a string using double quotes\n"

# variable with a text string using single quotes
txtString = 'Name'

print txtString
print '\nThe above is a string using single quotes\n'

# The following are examples of attribute queries for file based data
# e.g. shapefile, dBase, file geodatabase (feature class or table), or ArcSDE feature class

# NOTE: The field name must be encapsulated with double quotes for file-base feature classes
#       and tables.  The field name in this example is "Name"

# 1. This syntax is correct using double quotes

query = "\"Name\" = \'Downtown\'"

print query
print "\n1. The above example is an ArcGIS query with proper syntax using double quotes\n"

# 2. This syntax is incorrect using double quotes
query = "Name\" = \'Downtown\'"

print query
print '\n2. This example is an ArcGIS query, but the query syntax is incorrect \n' \
      '   because the field name, Name, is not properly bounded by doulbe quotes \n' \
      '   and the escape character.\n'

# 3. This syntax is correct using single quotes
query = '"Name" = \'Downtown\''

print query
print '\n3. This is an ArcGIS query, with the correct syntax using double quotes\n' \
      '   and the escape character.'
```

The script begins by showing a variable (txtString) set to a text string "Name". Two versions are shown representing the use of double and single quotes. In the next part of the script, a query variable (`query`) is assigned to a text string that would be used in the "where clause" parameter of a `SelectLayerByAttribute` or `SearchCursor` routine. Three different versions are illustrated if the query is used in file-based data (such as a shapefile, dBase table, file geodatabase feature class or table, or ArcSDE feature class or table). For file-based data double quotes are required to surround the field name. See Chapter 5 or the ArcGIS Help topic **Building a query expression** in ArcGIS that discusses proper format for different data types when creating query strings.

The general form of the query is "`<field_name>`" = `<value>` such that `Name` is the field name and `Downtown` is the name (value), for example, the name of a neighborhood.

The first query statement shows a properly formatted query string using double quotes to denote the character string. Note the placement of the escape character ("\") in the character string. The backslashes are put in to properly format the double and single quotes so that Python interprets the character string correctly so that it can properly be used in a subsequent ArcGIS process (such as `SelectLayerByAttribute` or a `SearchCursor`). The print statements that result from running the script are shown below.

In the second example, the character string is formatted correctly for Python so that it will not cause an error when using Check Module or Run. Note that the resulting print statement shows a missing double quote just before **Name**. This syntax will cause an error in the `SelectLayerByAttribute` or `SearchCursor`, because the query statement is not written correctly.

In the third example, the character string is formatted properly for an ArcGIS query string using single quotes that bounds the character string. Note that the escape character ("\") is used in slightly different locations. As can be shown above, the query string prints out with the proper format for use in a subsequent ArcGIS process.

The following script shows an example of a properly written query statement using single quotes and the escape character that is assigned to the variable query. The `query` variable is then used in the `SelectLayerByAttribute` routine.

```
try:

    #Check to see if feature layer exists
    #if it does, delete it

    if arcpy.Exists(nh_FLayer):
        arcpy.Delete_management(nh_FLayer)

    arcpy.MakeFeatureLayer_management(neighborhoods, nh_FLayer)

    query = '"Name" = \'Downtown\''
    print query

    arcpy.SelectLayerByAttribute_management(nh_FLayer, "NEW_SELECTION", query)

    result = arcpy.GetCount_management(nh_FLayer)
    print "Number of Selected Objects: " + str(result)

except:
```

When the results are printed, the following is shown in the Python Shell.

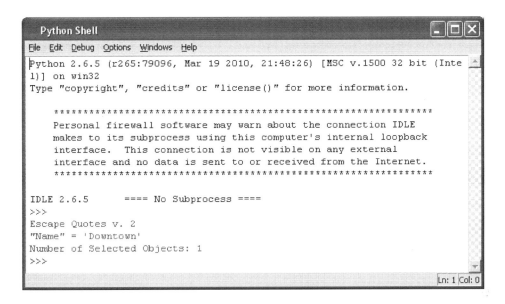

Being consistent with the use of single and double quotes and the escape characters are important considerations when developing code. Inconsistent use of quotes and the escape characters will often lead to code problems and errors

General Structure of a Python Script

A Python script is typically processed from the first line to the last line. A typical structure of a Python script is shown below and has the following general sections.

1. Title and Script Commentary/Notes
2. Import modules
3. Variable assignments (also Python function definitions)
4. Code Body
    a. Try block that contains the functional code of the script
    b. Except block – for error handling

As shown above, the script has a number of sections and help code developers organize their code development.

NOTE: This script can be used in the exercise below.

Title, Author, Date, and Script Comments

Typically, the first section of a Python script shows some comments that provide a title of the script as well as the author, date, dates of changes, etc. Comments can use a "#" or a series of three apostrophes ("'") that blocks out a "region" of code. This structure is commonly used to comment a section of code such as special notes about the script or blocks of code that an author wants to keep, but not process. Single line comments can be used to provide some in-line documentation for the script.

import Modules

With just a line or two, the required modules can be imported as shown above. The import modules must precede all other functioning lines of code.

Variable Definitions (and Python function definitions)

NOTE: Custom Python function definitions created by the code developer are often placed above any major section of Python script. Custom Python functions are not discussed in this book. The reader is encouraged to consult a Python text or the *python.org* website for more information.

The only requirement for variable definitions is that they be assigned before being used. Unlike some programming languages, variables do not have to be explicitly defined above the body of the script; however, defining some variables at the top of a script makes it more convenient for both the code developer and user to readily identify these variables if changes are required. Common examples include:

   a. Workspace paths
   b. Folder/Directory paths
   c. Variable name that represents a log file (e.g. to print statements to)

d. Input and Output data file, feature class, feature layer, table, or table view names

Code Body

The majority of the code will be written within the `try:` block. Any conditional statements, loops, and geoprocessing functions are typically written within the `try:` block. In addition, the `except:` block will contain any error messages that the code developer wishes to handle. Chapter 8 will discuss error handling in more detail.

Running a Python Script

Once a Python script has been written and saved, the script can be run to actually process the script. Often, after changes have been made to the script a three step process is used to run the script:

1. Save the script
2. Use the Check Module to check the script for any Python errors
3. Run the script

Check Module

Once edited code is saved, a code developer can check the Python script of Python syntax errors such as indentation, closed parentheses, missed semi-colons, misspelled key words, and quotation marks, among others. To do so, the developer can click **Run—Check Module** from the IDLE scripting window.

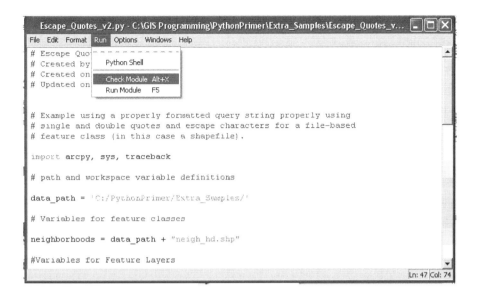

If errors appear, IDLE will highlight the line. All Python errors must be remedied before running a Python script; otherwise, the script will present an error message. The figure below shows an error message after running Check Module. The location in the line with the error is highlighted. This error will need to be fixed before the script can be "Run". It has been the experience of the author that most Python and ArcGIS related errors occur as a result of mistyping, improper Python syntax, and the proper parameters are not set to run ArcGIS functions.

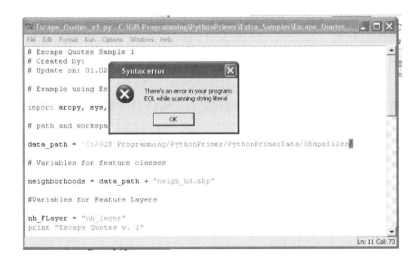

The developer should note that the Python IDLE script editor does not check for ArcGIS function syntax or perform function completion. Additional errors may occur after the check is successful. The developer will want to check ArcGIS function syntax, check to make sure variables are set to the proper values, and that query syntax is written correctly. Consulting ArcGIS Help for specific geoprocessing tool parameters, data type, syntax and examples are recommended. In addition, the developer may find it useful to test some of the geoprocessing functions using the ArcGIS Python Shell (sometimes noted as the Python Window in geoprocessing tool help documents).

Run Module

Once the Python script has been checked and/or fixed for Python syntax errors, the script can be run to process the lines of code. To do so, the developer can click **Run—Run Module** from the IDLE scripting window.

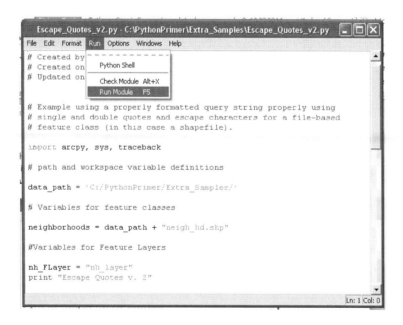

Handling Errors

If the code generates an error before successfully running the script, the developer will want to review the script again for syntax and logic problems, and then save and check the script again before re-running. Python will indicate the line of code the program fails. The developer should note that processing errors may occur before the line indicated in the Python error message. The figure below shows a Python script that passes the Check Module operation, but fails when the program runs. Note the parameter "OLD_SELECTION" highlighted below. "OLD_SELECTION" is not a selection type for SelectLayerByAttribute and hence an error message is produced.

NOTE: An error message is produced because the script contains code in the except: block that traces (checks for) ArcGIS problems. Without this exception block, the script would run without error, but the final print statement would not print out because the program does actually produces an error and stops running before the final print line (in the try: block).

```
nh_FLayer = "nh_layer"
print "Escape Quotes v. 2"

try:

    #Check to see if feature layer exists
    #if it does, delete it

    if arcpy.Exists(nh_FLayer):
        arcpy.Delete_management(nh_FLayer)

    arcpy.MakeFeatureLayer_management(neighborhoods, nh_FLayer)

    query = '"Name" = \'Downtown\''
    print query

    arcpy.SelectLayerByAttribute_management(nh_FLayer, "OLD_SELECTION", query)

    result = arcpy.GetCount_management(nh_FLayer)
    print "Number of Selected Objects: " + str(result)
```

If a Python script is run as a standalone script (without being run through a custom ArcGIS tool), error messages and print statements are "printed" to the Python Shell window. The figure below shows the error messages reported back from the script to the Python Shell. The specific error indicates that the function parameters are not valid and lists the valid types the *SelectLayerByAttribute* function accepts. The error message also indicates the specific line of the error. In addition, the specific ArcGIS Error is noted (ERROR 000800). A user can look this error up by doing a search on the error code in the ArcGIS Help. More details on ArcGIS errors are discussed in the next chapter.

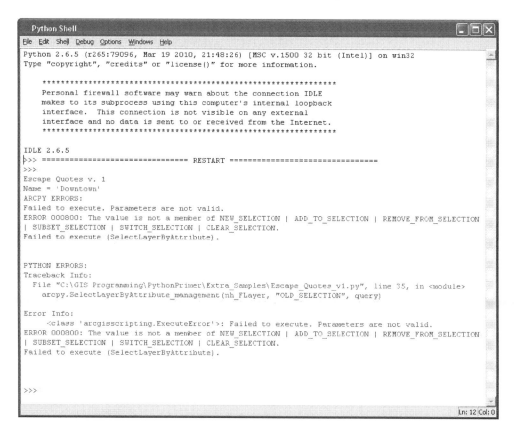

Summary

Chapter 3 provided an overview of the fundamental Python constructs that are often used with ArcGIS geoprocessing scripts. Many of these fundamentals will be used and referred to throughout the rest of the book. Readers are recommended to study and review these fundamentals and work through the script exercises for this chapter to gain more insight and familiarity with these Python concepts.

Exercise 3: Write a Simple Python Script

Python Concepts

Script layout and organization
`import` modules
Variable definitions
`try: except:` code blocks
`print` statement
Python IDLE functionality

This exercise demonstrates the overall organization of a Python script and some of the methods to create variables. There are no ArcGIS functions used in this script. Create a simple script with the following conditions. Answer the following questions below.

NOTE: Use the **\PythonPrimer\Chapter03\General_Layout.py** script to begin writing the script below; otherwise, open a blank Python scripting window and add the following syntax. Use the text as needed to help write the script.

NOTE: If the reader does a copy/paste of the electronic version of the text, the quotes may need to be retyped within the Python scripting editor.

Open a new or the General_Layout.py script

1. Start the Python IDLE editor from **Start—All Programs—ArcGIS—Python2.6—IDLE**
2. Click **File—New Window** to open a blank Python scripting window
3. From the Python IDLE window click **File—Open** and browse for the **General_Layout.py** script file. (Alternatively, the reader can browse the Windows Explorer and go to the **Chapter03** folder and right click on the **General_Layout.py** script and choose **Edit with IDLE**.

Start writing the script

4. Add a title, name, creation date, and notes section
5. Add the `import os` and `traceback` modules

6. Create the following variables:

    a. String variable assigned to the following text.

    ```
    'I can't wait to start programming with Python and ArcGIS!'
    ```

    b. A number variable, x = 1

7. Create a `try:` block and add the following. Make sure to properly indent (usually one tab). The `if` statement will be indented because it will be placed inside the try: block. The print statements in the "if" block will be indented. See the text above and consult a Python text or resource if necessary. The first print statement will be written on a single line.

    ```
    if x == 1:
        print 'This is my ' + str(x) + 'st Python script \n' + <put your string variable name here>
        print 'Program Successful!'
    ```

    Make sure to replace the `<put your string variable here>` (including the <>) with the name of your variable.

8. Add an `except:` block and add the following code. Make sure to properly indent.

    ```
    print 'Program failed.'
    print 'z is not defined.  Assign "z" a value and re-run the script.'
    ```

9. Add a comment line for each statement in the `try:` and `except:` block.
10. Above the `try:` block add the following statement

    ```
    print 'Starting program...'
    ```

11. Save the script (make sure to save the file as a *.py* file name extension).
12. Check the script for any errors by going to the script window where the code was typed into and click **Run→ Check Module**. If any errors appear,

resolve them. Most of the errors at this point should be related to typos, missed quotes, variable names, and indentation. See above for comments on using the Check Module.

For this program, search through the code and make sure the syntax is typed correctly. If needed, the solution can be consulted, but get in the habit of attempting to "troubleshoot" the code on your own and consulting a number of the resources listed in the text.

13. After any errors are resolved, click **Run→Run Module** from the script editing window. Review the results in the Python Shell.

Make the following change to the code, check the script, and run again.

14. Add the following line within the "if" statement (i.e. indented within the if x == 1: block) after the first print statement and then re-run the script. Notice what happens in the script.

    ```
    print z
    ```

Chapter 3: Questions

1. What are the following used for?

    a. Python IDLE
    b. Python Shell

2. What are two key syntax elements that are important standards to follow when writing Python code?

3. What are variables and how are they used?

4. What is a good recommendation for creating variable names?

5. Give an example for each of the variable types:

    a. String variable
    b. Number variable

6. What are the conditional statements and looping structures typically used in Python?

7. What are the `try:` and `except:` blocks used for?

8. Briefly describe what the Check Module and Run Module function do in Python IDLE.

The following questions focus on the script created in Exercise 3.

1. Where does the escape character need to be placed?

2. What happens if the escape character is not used in the string variable? Hint: Try running the script by removing the escape character.

3. In the print statement above, what does the "\n" represent?

What does it do when the script is run? Do a search on *Python.org* or consult a Python text (such as *Learning Python*).

4. When `print z` is added, describe what occurs in the script?

Chapter 4   Writing a Basic Geoprocessing Python Script

Overview

This chapter will expand on the Python fundamentals for Chapter 3 and introduce more ArcGIS specific concepts, organization, and syntax. The latter part of the chapter will focus on setting up workspaces, variables, and the Clip and Buffer geoprocessing routines which will be used to demonstrate writing the first geoprocessing script using Python and ArcGIS.

Getting Ready to Create an ArcGIS Geoprocessing Python Script

Before starting a Python script the code developer should consider what kind of geoprocessing tasks will be accomplished and how the script will be organized. For example, if a series of geoprocessing tasks have already been identified and performed manually, a list of the tasks and the order in which they are implemented can serve as the basic outline for the Python script. If, on the other hand, a code developer is not sure of the geoprocessing tasks, then the developer should consider consulting the ArcGIS help and performing some of the individual tasks manually in ArcMap to research and discover the appropriate geoprocesses and order in which they need to be organized. In addition, the code developer should consider developing the geoprocessing workflows described in the Introduction as well as using ModelBuilder as described in Chapter 2 to assist in the code development process.

Using Pseudo-code to Outline Geoprocessing Tasks

A common method for outlining coding tasks is the use of "pseudo-code." Pseudo-code is no more than using comments to draw up an outline of coding tasks in plain English. Similar to developing an outline for a paper, book, or written document, pseudo-code begins with the broad general tasks and then becomes refined to include more specific tasks, ideas to test, and notes for further consideration. Code development can take weeks or months to create, refine, and test, so having an in-line set of documentation can be very helpful to make progress.

The following example script does not have much actual code, but is a good start at an outline for the general coding tasks.

```
# Demo 2: Pseduo-code for geoprocessing

# Created by:  Nathan Jennings
# Created on:  11.26.2010
# Updated on:  11.26.2010

'''
NOTES:  This section is reserved for general notes about the script
'''


# import the required modules for the python script
import arcpy, os, traceback


# define common variables here

try:

    # Create feature layer


    # Select Features by Attribute
    # Query:  Select all land use types that are single family residen


    # Compute land use value


    # Research parcel data to determine proper value for multiplier

    # Add new field for land use value
    # Calculate land use value  Structure Value * 0.0125

    # Copy all rows to an output table
    # Use Copy Rows function


except:
```

Before beginning code development some of the elements specific to ArcGIS geoprocessing need further explanation.

arcpy Module Overview

The `arcpy` module is the required module to perform ArcGIS geoprocessing tasks using Python and allows reference to any of the available geoprocessing methods and properties provided by Esri.

Examples of geoprocessing methods are any of the geoprocessing routines found in the ArcToolbox. For example,

> Clip to clip features
> Buffer to buffer features
> Create Featureclass to create feature classes
> Make Feature Layer or Make Table View to create feature layers or table views
> Add Field or Delete Field to add to or delete fields from a table

Examples of geoprocessing properties usually refer to specific elements of feature classes, tables, images, and spatial reference parameters among others. Some properties include:

> Field names
> Data types (points, lines, polygons)
> Spatial extent parameters such as the coordinate system
> Number of bands in an image

The `arcpy` module is used in a similar way as the pre-ArcGIS 10 `arcgisscripting` module, however, the developer does not need to create a geoprocessing object, but rather use the `arcpy` module directly. The `import arcpy` module must be included in the script before any reference to and use of arcpy methods and properties.

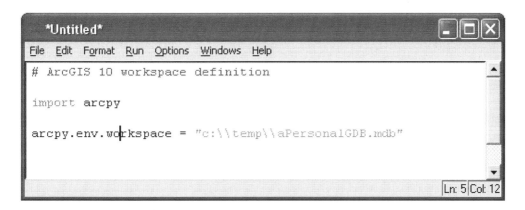

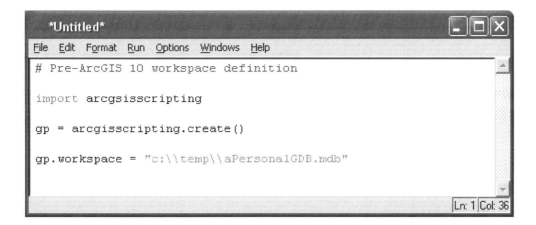

Alternatively, specific import modules can be referenced and used in Python such as the `os` and `sys` modules. This can simplify code, but the developer needs to be aware of how the modules are referenced, since some confusion can occur and/or syntax can be more difficult to read. See the **ArcGIS Web Help** under **Importing ArcPy**.

The script in the figure below shows one method of importing all of the `arcpy` modules and all of the mapping routines that can be used for working with ArcMap documents. The use of `from` and `import *` can simplify some of the code writing without having to write `arcpy.mapping` in front of each mapping routine such as `MapDocument` and `ListDataFrames` routines.

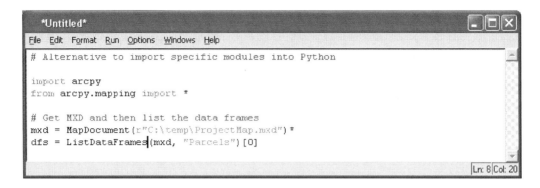

Note the use of the "r" delimiter and the single backslash in the above example to process the data path as raw string as opposed to the double backslash that does not require the "r" delimiter. See Chapter 3 for more details.

Alternatively, the following can be written without the use of `from arcpy.mapping import *`. The difference here is the full reference to the `arcpy.mapping` module is required for all `arcpy.mapping` routines (e.g. `MapDocument` and `ListDataFrames`). See figure below.

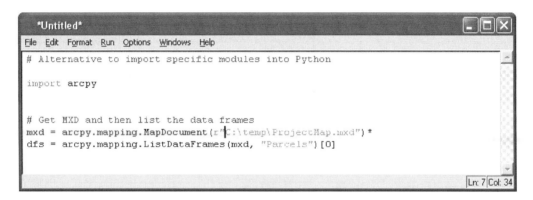

The reader will see a variety of import options within the ArcGIS Help and sample code. The most commonly used import modules are:

```
import arcpy, os, sys, traceback
```

It is recommended that the reader review the ArcGIS geoprocessing routines to understand how ArcGIS uses the different coding strategies.

Workspace and Data Path Variables

Workspace and data paths can be used to point to a specific location on a computer or server disk for data. Data paths can be used to set directory locations, personal or file geodatabase or SDE geodatabase locations. The data paths are simply character strings in Python. The main difference between defining a variable to a data path (i.e. a set of characters representing the path to data) versus an ArcGIS workspace, is the workspace contains additional properties that can be accessed once a workspace has been defined (such as the type of workspace – file or personal geodatabase, folder, or ArcSDE geodatabase). In ArcGIS 10 and `arcpy`, the workspace is accessed through the environment class (`arcpy.env`).

NOTE: A class is a collection of programming routines with related function.

The code snippets in the figure below show examples of each kind of variable definition.

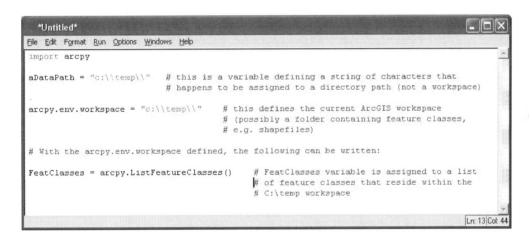

A list of feature classes cannot be obtained using the `aDataPath` variable above because it is not defined as a workspace (i.e. `arcpy.env.workspace` is not used to obtain the current workspace).

See also the ArcGIS Help documentation topic **Using Environment Settings in Python and Setting Paths to Data in Python**.

The figure below shows some examples of geodatabase workspace variables. Notice that a variable is set to accept the geodatabase connection. The above examples did not set a variable to a workspace. Also notice that a variable *outpath* is set to a separate ArcSDE connection and then another variable (*tree_fc*) is set to a feature class that uses the output variable.

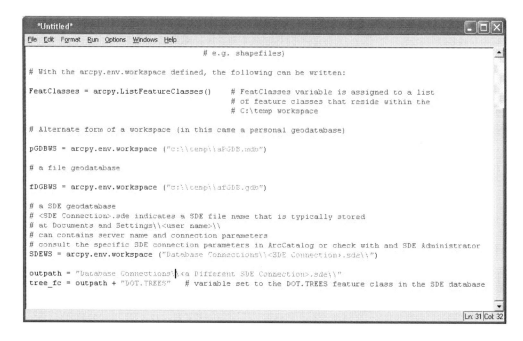

For those using network connections to store data, relative paths (also UNC – Universal Network Connection) can also be used (see figure below). For local drive mapping, relative paths cannot be used.

Define Variables

While developing a Python script for geoprocessing, the programmer will need to review the kinds of geoprocessing functions to be implemented in the code and determine the kinds of variables that might be required. Typically, "hard coded" values used as parameters, such as specific file names, specific query strings, or function parameters make code less flexible and require the code to be changed with different data. Variables will help make code more flexible and can be used with different kinds of data for the same purposes. Usually variables will be defined for workspaces, input and output data, names for query strings, and geoprocessing function parameters among others. Once a variable is defined, it can be used multiple times throughout a script.

Hard Coded Parameters

The Clip function below does not use variables, but rather has specific values "hard coded" for each parameter. The program must use the specific quoted values to process the Clip function. If the data path or data change, then changes to these parameters will need to be changed. Refer to the `hard_coded.py` script that can be found in the **\PythonPrimer\Extra_Samples** folder.

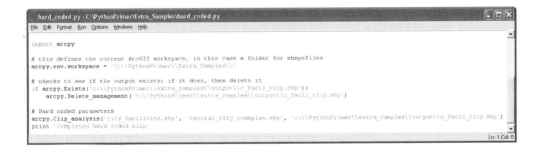

Parameters Using Variables

The same Clip function below uses variables instead of the hard coded values shown above. Even though specific variables are "hard coded," only the variable assignments need to be changed (typically only one time) rather than each occurrence of a specific value throughout the script. In later chapters the reader will discover how the variables can be dynamically assigned by user input from a custom ArcGIS tool, especially when the geoprocessing will be performed in an automated fashion. Using variables for the function parameters make this possible.

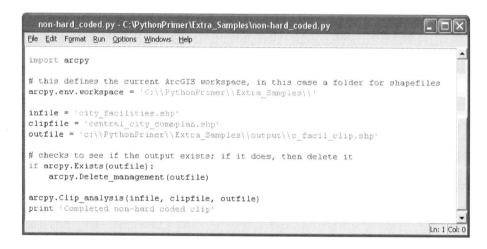

Add and Modify Geoprocessing Functions

Once an ArcGIS Python script has been started, other variables and geoprocessing functions can be added. The user should consult ArcGIS Help documentation or other sources to learn about the specific geoprocessing function requirements, parameters, and parameter formats. Additional research may be required to discover the function (or functions) and the pre-requisites for unfamiliar geoprocessing functions. Typically, this involves researching ArcGIS Help, Esri user forums, and Internet searches regarding Esri ArcGIS tasks and processes. Since this chapter introduces writing a basic geoprocessing script, the following section outlines some of the steps a code developer can use to learn about a geoprocessing function before adding it to a Python script.

Search ArcGIS Help

In this example, the Buffer function will be added to a Python script. Before adding the function, the code developer can search the ArcGIS Help on the Buffer function and learn how it operates. The developer can open the ArcGIS Help and type in Buffer. Alternatively, if the user is familiar with the Buffer tool, the tool can be opened from the ArcGIS Toolbox and then open the tool help.

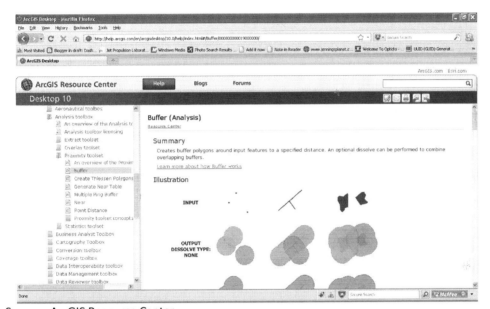

Source: ArcGIS Resource Center,
http://help.arcgis.com/en/arcgisdesktop/10.0/help/index.html#//000800000019000000.htm

If the programmer is unfamiliar with this function, the help should be read and reviewed to better understand what the function accomplishes. Toward the bottom of the Buffer Help document the general syntax for the function is shown with some descriptions of each parameter.

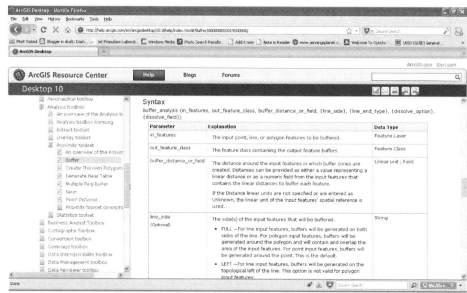

Source: ArcGIS Resource Center,
http://help.arcgis.com/en/arcgisdesktop/10.0/help/index.html#//000800000019000000.htm

In addition, the ArcGIS Help often has an example that shows most of the geoprocessing syntax in Python scripting format.

Source: ArcGIS Resource Center,
http://help.arcgis.com/en/arcgisdesktop/10.0/help/index.html#//000800000019000000.htm

This documentation can provide insight for the programmer to properly write the syntax for a given geoprocessing function as well as provide some clues to other elements related to the function (for instance optional parameters and other conditions such as feature layers or table views. Feature layers and table views will be discussed in Chapter 5).

Once the documentation is reviewed, the programmer can copy or type the syntax to a Python editor for the given function as well as adding and defining any additional variables required for the function. To copy code from an ArcGIS Help document, the developer can highlight the line (or lines) of the example script, right-click and select **Copy** from the drop down list (or use Ctrl + C keys) and then Paste the selected code into an open Python script using the Python IDLE editor **Edit—Paste** operation (or Ctrl + V). See figures below.

NOTE: The code developer will want to check the syntax being copied from ArcGIS Help, since sometimes the characters can change (for example, double quotes).

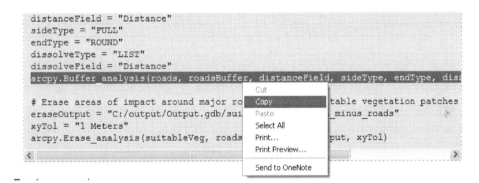

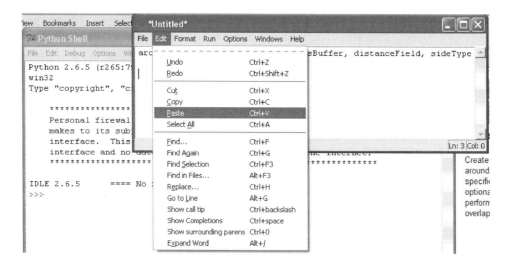

After the syntax is pasted into the Python editor, specific changes can be made to the individual parameters such as using variables defined for the script being developed by the programmer. The figure below shows the Buffer syntax added to an existing Python script. Notice the code added from the ArcGIS Buffer Help has been commented out since it is used for reference in the script. In addition, an additional variable `out_buffer` has been defined and used in the Buffer routine used below. Some additional comments have been added to the script to clarify some of the changes and modifications. This script can be found at **\PythonPrimer\Chapter02\Clip_Data2.py**.

```
import arcpy

# Local variables:
Sacramento_Streets = "Sacramento_Streets"
Downtown = "Downtown"
Clip_Results2_shp = "C:\\PythonPrimer\\Chapter02\\Data\\Clip_Results2.shp"

out_buffer = "c:\\PythonPrimer\\Chapter02\\Data\\out_buffer.shp"

# Process: Clip
arcpy.Clip_analysis(Sacramento_Streets, Downtown, Clip_Results2_shp, "")

# Inserted (Pasted) code from ArcGIS Help --- [Commented code below]
# Modifications have been made to show changes specific to this script.

# arcpy.Buffer_analysis(roads, roadsBuffer, distanceField, sideType, endType, dissolveType, dissolveField)

arcpy.Buffer_analysis(Clip_Results2_shp, roadsBuffer, "100 FEET", "", "")
```

ArcGIS Toolbox Aliases

ArcGIS tools (i.e. those found in the ArcToolbox) are organized into toolboxes that share related activities. For example, tools that are used for creating feature classes, working with attribute fields, and creating feature layers and table views can be found under the Data Management Tools toolbox. When using these tools in Python, it is good practice to use the alias toolbox name that is associated with the specific tool (i.e. the toolbox alias indicate the specific ArcGIS toolbox where the tool can be found). The general format of this syntax is:

```
arcpy.ArcGISTool_toolalias(<parameters for tool>)
```

For example, the Clip tool has the following syntax.

```
arcpy.Clip_analysis(<input feature class>, <clip feature class>, <output feature class>, {tolerance}, {units})
```

Notice that the Clip tool can be found in the Analysis toolbox.

The specific toolbox aliases can be found in the ArcGIS Help topic **Geoprocessing—Managing toolboxes—Renaming a toolbox: name, label, and alias**. An alternative method for writing ArcGIS Tool syntax can be found here: **Geoprocessing—Geoprocessing with Python—Accessing tools—Using tools in Python** (see the ArcGIS section on Tool organization). Instead of using the toolbox alias, a code developer can use the module form.

```
arcpy.toolboxname.tool(<parameters>)
```

For the Clip tool shown above, the syntax would be

```
arcpy.analysis.Clip(<input feature class>, <clip feature class>, <output feature class>, {tolerance}, {units})
```

Syntax for the tool must also be maintained (such as capitalization of characters; see the specific tool help for the exact syntax). The method used to access ArcGIS tools is determined by personal preference; however, one of the two methods must be used when accessing ArcGIS tools. The code developer should adopt one of the two methods and use them consistently in Python scripts.

Summary

Chapter 4 introduced some of the specific ArcGIS Python concepts that are required to begin writing useful Python code to perform geoprocessing. The reader was introduced to the use of workspaces, data paths, and variables to set up data and ArcGIS tool parameters versus hard coding these parameters. Doing so provides the fundamentals to write flexible and extended Python scripts for multiple tasks, many of which will be covered in later chapters.

Demo 4: Writing a Clip Features Script

This demonstration combines some of the concepts mentioned throughout this chapter. After completing this demo the reader should be able to understand the following.

ArcGIS Concepts

Import the `arcpy` module
Define a workspace
Refer to ArcGIS help for the Clip geoprocessing routine
Successfully run the Clip routine

Python Concepts

Create a pseudo-code outline for the script
Create some variables for use in the script
Set up `try` and `except` blocks
Add some error handling text to the except block

The reader should attempt to write the code on their own to gain practice in developing programming skills. The Demo 4 solution is provided and can be consulted. Exercise 4 will need to use the code developed in Demo 4.

1. Create Pseudo-code

    a. Start a new Python script window by opening IDLE
    b. Name the script **Clip_Features.py** and save it to the **\MyData** folder.
    c. Add the following pseudo-code information to the Python script editor. Add your name and the current date at the top of the script. The script up to this point should look like the following:

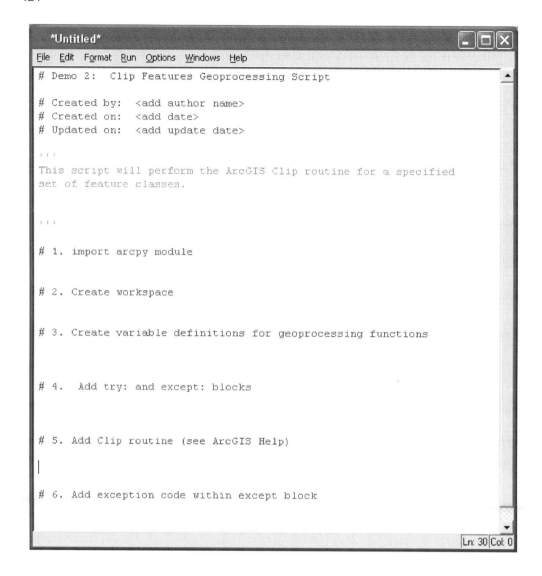

2. Start building the script.

   a. Add the following information.

   Add additional commentary
   Import `arcpy` module (and the `sys` and `traceback` modules, see below)
   Add a workspace

If needed, change the path for the workspace to the folder that contains the **\PythonPrimer\Chapter04\Data** folder. NOTE: The workspace path may need to be changed depending on where the reader placed the data for the demo/exercise. After adding the information above, the script should look similar to the figure below.

3. Consult ArcGIS Tool Help for the Clip routine.

Consult the ArcGIS Tool Help for the Clip geoprocessing function. ArcMap or ArcCatalog can be used to locate the tools. Clip is under **Analysis Tools—Extract—Clip**. Alternatively, use the Search Tab in ArcToolbox and type in each of the geoprocessing functions to locate the tool. Click on the Tool Help to review the specific documentation. Make sure to review the description of the function as well as the required and optional parameters. Scroll to the bottom of the help to see a Python script example using the tool.

The following parameters will be required for the geoprocessing functions in the demo script. Refer to the **\PythonPrimer\Chapter04\Data** folder to find the data. Use **\PythonPrimer\Chapter04\MyData** to store any output files. Load it into ArcMap to view it, to see what the data looks like.

Clip Parameters

Input features (**City_Facilities.shp** - City Facilities); **City_Facilities.lyr** exists that contains some standard symbology. The layer file is not required for the script to function, but can be used in ArcMap for viewing.

Clip features (**Central_City_CommPlan.shp** - Central City Community Plan Boundary)

Output features – a feature class name, **\PythonPrimer\Chapter04\MyData\City_Facilities_Clip.shp**. The reader can change the path to match a location on their local system.

a. Create variables for each of the above required parameters. After creating variables for the **Clip** geoprocess, the script should look similar to the following:

```
# Demo 4:   Clip Features Geoprocessing Script

# Created by:   Nathan Jennings
#               www.jenningsplanet.com
# Created on:   11.26.2010
# Updated on:   10.29.2011
# Copyright:    2011

'''
This script will perform the ArcGIS Clip routine for a specified
set of feature classes.

This script demonstrates:

1. Workspaces
2. Input and output feature classes
3. Use of ArGIS geoprocessing functions
4. Defining paraeters for the ArcGIS Clip routine programmatically
5. try: except: for error handling

'''

# 1. import arcpy module

import arcpy, sys, traceback

# 2. Create workspace

arcpy.env.workspace = "C:\\PythonPrimer\\Chapter04\\Data\\"

# 3. Create variable definitions for geoprocessing functions

outpath = "C:\\PythonPrimer\\Chapter04\\Data\\"

infile = "City_Facilities.shp"
clipfile = "Central_City_CommPlan.shp"
outfile = outpath + "City_Facilities_Clip.shp"

# 4.   Add try: and except: blocks
```

4. Add **Clip** routine and additional ArcGIS code

Type in the following syntax for the **Clip** geoprocessing function (see figure below). Make sure to use the toolbox alias name or the alternative method described above. Add a couple of `print` statements that can print out to the Python Shell to monitor the script's progress. Note how the variables are inserted into the respective geoprocessing script and how the statements are indented because they are inside of the `try:` block.

In addition, add the two lines to check to see if the output file exists (i.e. see the "if" statement with `arcpy.Exists` and `arcpy.Delete_management` functions). This will be discussed in subsequent chapters. Notice that `arcpy.Exists` does not have an associated toolbox alias, since `Exists` is an ArcGIS function and not a tool. Typically, the ArcGIS functions that are not associated with tools typically perform general tasks such as checking for the existence of data. See the ArcGIS Help: **Geoprocessing—Geoprocessing with Python—Accessing tools—Using functions in Python** and **Geoprocessing—The ArcPy site package—Functions—Alphabetical list of ArcPy functions**.

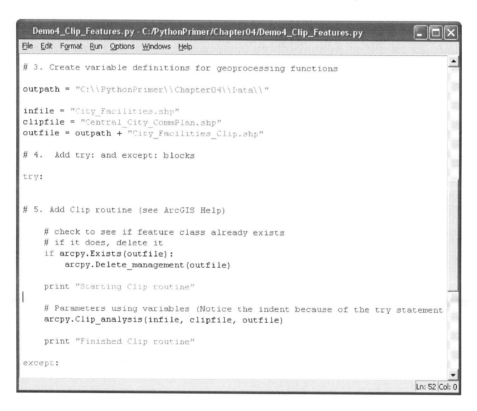

5. Add Exception Code

Add the following exception code to the script. Go to the **\PythonPrimer\Chapter04** folder and open the **Exception.py** script in Python IDLE. Select all of the text, copy, and then paste it into the Clip Python script being developed for this demo. Use the Python editor **Edit—Copy** (or Ctrl + C keys) and then the **Edit--Paste** (or Ctrl + V) from the Python script editor (of the Clip script being developed) to paste the code within the script. Make sure to place the cursor within the `except :` block before pasting the code. Make sure to indent the exception code as shown below. If the exception code is not indented properly by default, highlight all of the exception text inside the Python script editor, click the **Format—Indent Region** option within the Python script editor. This will indent the entire block of selected code. Note that some lines may extend beyond the window display. This is ok.

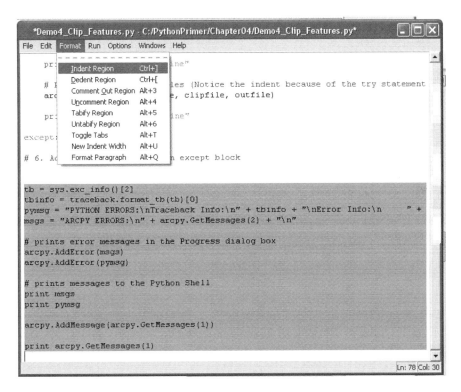

The text below shows the indented code block.

```
    print "Starting Clip routine"

    # Parameters using variables (Notice the indent because of the try statement
    arcpy.Clip_analysis(infile, clipfile, outfile)

    print "Finished Clip routine"
except:
# 6. Add exception code within except block

    # Notice the indent because of the except block
    tb = sys.exc_info()[2]
    tbinfo = traceback.format_tb(tb)[0]
    pymsg = "PYTHON ERRORS:\nTraceback Info:\n" + tbinfo + "\nError Info:\n
    msgs = "ARCPY ERRORS:\n" + arcpy.GetMessages(2) + "\n"

    # prints error messages in the Progress dialog box
    arcpy.AddError(msgs)
    arcpy.AddError(pymsg)

    # prints messages to the Python Shell
    print msgs
    print pymsg

    arcpy.AddMessage(arcpy.GetMessages(1))

    print arcpy.GetMessages(1)
```

Also add the `sys` and `traceback` modules to the import line of the script. Methods from these modules are used in the exception code block. If these modules are not imported and the script incurs an error, the Python script will result in an error message indicating that a name is not defined.

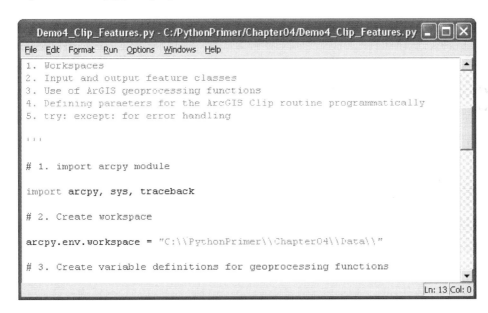

The import line in the script should show the following:

```
import arcpy, sys, traceback
```

6. Click File—Save.

7. Check the script for Python errors

Click **Run—Check Module** to check for any errors. If the script was written correctly, no error messages should pop up and the Python Shell prompt (>>>) will appear.

If errors appear, try and figure them out or consult the **Demo4_Clip_Features.py** script in the **\PythonPrimer\Chapter04** folder. Once typos and other syntax issues are resolved click save. Click Run—Check Module to make sure the script does not have any errors.

8. Run the script

Click **Run—Run Module**.

The script will run and the two print statements will appear in the Python Shell script. If errors occur, note them, and consult the **Demo4_Clip_Features.py** script to remedy any issues. If errors are found, fix them, save the script and re-run the script.

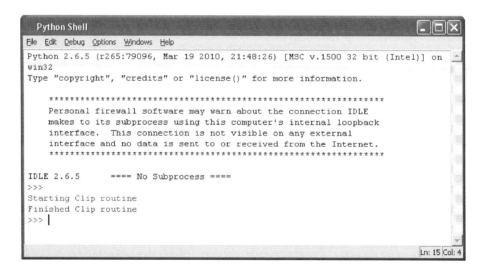

9. Review results in ArcMap

Close the Python script and the Python Shell window. This will release any data locks on the files so they can be properly seen and used in ArcMap or ArcCatalog.

Open ArcMap and add the **City_Facilities.shp, Central_City_CommPlan.shp,** (if not already added) and the **City_Facilities_Clip.shp**.

Notice the City Facilities shapefile has been clipped (dots) with the Central City Community Plan boundary.

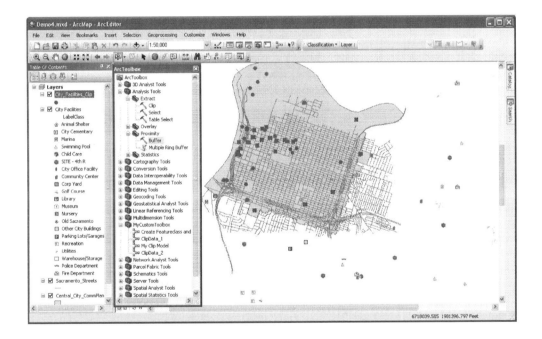

Exercise 4: Add the Buffer Routine to the Clip Features Script

Using the code developed in **Demo 4**, do the following to add the **Buffer** routine to the script after the **Clip** routine. Review the ArcGIS Help for the Buffer routine.

1. Add variables for the following required buffer routine variables:

    a. **Input features** for the buffer routine will use the variable created for the output clip feature class. (The `outfile` variable used in the Clip routine in Demo 4 is the "input features for the Buffer routine in Exercise 4).

    b. **Output buffer features** – a feature class name that will represent the "buffered" features (i.e.
    \PythonPrimer\Chapter04\MyData\City_Facilities_Clip_Buffer.shp).

    c. **Buffer distance** – a variable representing a number (e.g. 10, 100, 1000, etc.).

    d. **Buffer units** – a variable representing the word "Feet" or "Meters". This will represent the unit portion of the buffer distance parameter. The buffer distance consists of two parts, a unit measure (10, 100, 1000, etc) and the unit type (feet, meters, miles, etc). The format for the parameter will look like this: "100 Feet", "1000 Meters", etc.

    NOTE: To use the Buffer distance and the Buffer units together, the code developer will need to "concatenate" the two variables. To obtain the format shown above for the buffer distance, the following syntax will be needed: **str(Buffer_distance) + " " + Buffer_units**. *str()* converts the number to a "string" type so that it can be "concatenated" with the word "Feet." Python cannot concatenate a number with a string. The double quote has a space between each quote. This represents the space between the number and the unit type, thus if Buffer_distance is 100 and Buffer_units is "Feet", then the above syntax will represent "100 Feet".

The optional parameters are not needed for the exercise and do not need to be added.

2. Add other print statements if desired.

3. Add an `if` statement similar to the example above that checks to see if the output buffer file exists; if it does, then delete it. This will be placed after the Clip routine and before the Buffer routine. *Remember that a different variable will be used for this if statement that points to the "buffer output features" feature class. (See Demo 4, step 4).*

4. Make sure to Save and "check" the module before running. Save the script to a new name **(\MyData\City_Facilities_Clip_Buffer.py)**. Attempt to fix any syntax problems.

5. Run the script to see if the clip and the buffer works.

6. After successfully running the Python script, close the Python script editor and the Python Shell and then review the results in ArcMap.

ArcMap should show the clipped and buffered data like the following:

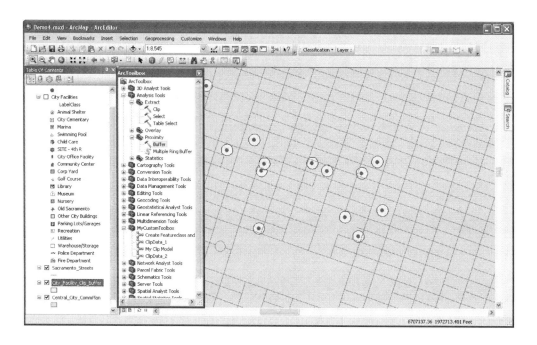

Chapter 4: Questions

The following questions focus on the Chapter 4 material.

1. What is the benefit of writing "pseudo-code"?

2. Where in a Python script do the `import` modules need to exist?

3. What is the main difference between setting a variable to a workspace versus setting a variable to a data path?

4. What does "hard coded" refer to? Describe.

5. What is the benefit of using variables as parameters?

6. What is the suggested location of setting variables? Why does this make sense?

7. If a code developer does not know if an ArcGIS tool, function, or geoprocess exists, what are two options can one use to find the tool, function, or geoprocess and learn about its parameters?

The following questions focus on the script created in Exercise 4.

1. If the `outpath` variable (for the Clip Features script) was not used in the script, what would the `outfile` variable be assigned to? Write the syntax.

2. If the workspace environment was not set, would the script be able to function properly without it? Why or why not. (Hint: Try commenting out the arcpy environment line and re-running the Python script). Describe what happens and provide a screen shot of the results.

3. Describe what the `arcpy.Exists` and `arcpy.Delete_management` functions do in the script.

4. How do you know the `sys` and `traceback` modules are required for this script? Hint: Look through the entire code written for this exercise.

5. What happens if you take out the `sys` module and re-check the script? Provide a screen shot of the results in the Python Shell.

6. What happens if you take out the `traceback` module and re-run the script? Provide a screen shot of the results in the Python Shell.

7. For the buffer script, write the syntax used to set a variable to the output buffer file.

8. Why do the `buffer_distance` and `buffer_units` variables need to be combined in order to work properly in the Buffer routine?

Section II: Writing Python Scripts for Common Geoprocessing Tasks

Section I described the Python fundamentals and syntax as well as the general layout and organization of Python scripts. In addition, Section I demonstrated the construction of simple Python scripts from some familiar ArcGIS geoprocessing tools inserting user defined variables that were then required as the geoprocessing tool parameters.

Section II focuses on the most commonly used geoprocessing tasks implemented within ArcGIS for different kinds of data and geoprocessing analyses. Section II comprises the bulk of *A Python Primer for ArcGIS* and will discuss the general Python and ArcGIS methods, syntax, and practical workflows for implementing these common tasks.

The major geoprocesses covered in Section II are:

Chapter 5 – Querying and selecting data using the *MakeFeature, TableView, SelectLayerByAttribute, and SelectLayerByLocation* routines.
Chapter 6 – Creating and using cursors, table joins and the `for` and `while` loops.
Chapter 7 – Operating on lists and describing data and the `for` loop.
Chapter 8 – Custom error handling, writing messages to log files, and being able to send messages to the ArcToolbox message dialog box.
Chapter 9 – Introduces and describes the `mapping` module to create maps and automate map production.

Chapter 5   Querying and Selecting Data

Overview

A common operation in GIS is to query data by selecting specified geographic features or rows in a table and then doing something with the selected data (for example, a geoprocess). Querying and selecting data can be performed on one or more attributes ("attribute query") or through spatial overlay processes ("spatial query"). Both kinds of queries can be performed using ArcGIS tools and Python. A number of spatial overlay processes exist in ArcGIS, such as Clip, Buffer, Intersect, and Union, however, this section covers two querying processes:

1. Select by Attribute
2. Select by Location

Using both kinds of queries in GIS allow an analyst to ask questions of the data both in a tabular fashion as well as based on spatial coincidence. An example of an attribute query might be "Select all of the cities in a cities point file that have a population greater than 100,000." An example of a spatial query might be, "Select all of the parcels that fall inside of a flood zone." In both cases, the analyst is making an inquiry of the data that meets one or more conditions. The examples shown above represent simple queries, however, in practice queries can be complex with multiple conditions.

Often, when an ArcGSIS user needs to perform an attribute or spatial query, one of the following Selection menu options are used.

For geoprocessing and model building the actual ArcGIS tools exist in the **Data Management Toolbox** within the **Layers and Table Views Toolset**. These are the tools that will be used in Python scripts.

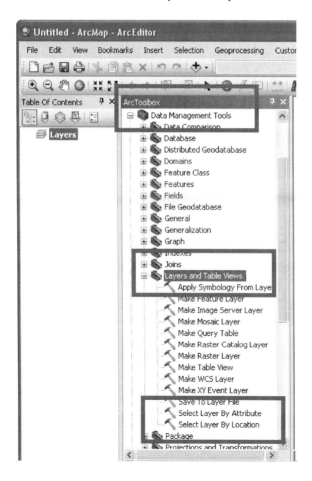

To review the functionality of each tool, the respective tool dialog box is shown below.

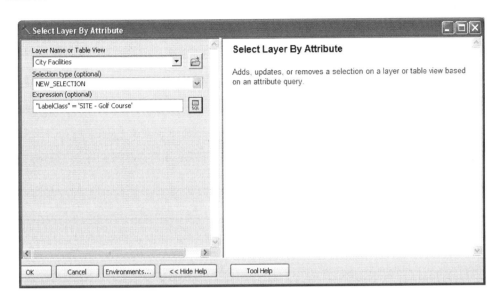

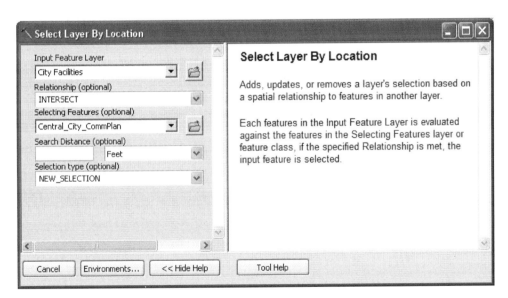

The general structure of each selection type contains certain parameters that must be provided by the user. When developing Python scripts the required parameters must be added whereas the optional parameters can be left out or used when needed. Any parameter that is optional and indicates a "default"

value (such as the selection type) can be left out unless a different option is chosen. The following code snippets show examples of using and not using the optional parameters for the SelectLayerByAttribute and SelectLayerByLocation routines.

```
# Select layer by attribute without the optional query parameter
# Assume the feat_layer variable is properly defined)
arcpy.SelectLayerByAttribute_management(feat_layer, "NEW_SELECTION")

# Select features by attribute using the optional query
query = '"CLASS" = \'H\''
arcpy.SelectLayerByAttribute_management(feat_layer, "NEW_SELECTION", query)

# Select layer by location without using the following optional parameter:
# search distance
arcpy.SelectLayerByLocation_management(poly_layer, "INTERSECT", \
                                        feat_layer, "", "NEW_SELECTION")

# Select layer by location using all of the optional parameters:
# Assume poly_layer, feat_layer, and search_distance are properly defined
arcpy.SelectLayerByLocation_management(poly_layer, "INTERSECT", \
                                        feat_layer, search_distance, "NEW_SELECTION")
```

The first line shows the SelectLayerByAttribute routine using the required feature layer and the optional selection type (i.e. "NEW_SELECTION"). The other optional parameter (query statement) is not shown. When this line is processed by Python all of the features will be selected as a new selection because no query statement is provided to limit the selected features. In the second SelectLayerByAttribute line, a query statement is created and used in the routine. In this case, only the roads with a CLASS = 'H' will be selected, thus limiting the selected features.

In the first SelectLayerBylocation routine the optional search distance parameter is left out. Note that the other optional parameters are added. A place holder for the search distance is provided (noted with ""). If an optional parameter is not used and is not a the end of the parameter list for the routine, a placeholder of double quotes ("") or single quotes ('') must be used. The order of the parameters must be maintained to process the routine correctly. In the second SelectLayerByLocation routine, all of the parameters are filled in with either variables or key words for a the relationship and selection types.

Prerequisites

Before developing Python code for selecting features, the reader needs to understand how to construct a query expression that can be used in Python script as well as the differences between "feature classes" or "tables" and "feature layers" or "table views." Query syntax for Python scripting follows the same syntax rules when constructing a query in ArcMap. The code developer needs to be aware of the data format being used in the geoprocess as well as the values and the Python syntax requirements that make up the conditions of the query. In some cases, special characters such as the Python escape character ("\"), a wildcard character (e.g. % or *), or a mathematical expression may be needed. Designing query syntax can take the code developer considerable time, especially when the queries are complex. Another challenging area of developing Python scripts for ArcGIS is the use of feature layers and table views. Within ArcGIS, these kinds of operations are done automatically when the user adds data or creates a new data set. In Python, the code developer needs to explicitly write the "instruction" to create the feature layers or table views. Being able to construct and troubleshoot query syntax as well as to identify when feature layers or table views are required and writing the code to perform these activities is the focus of this chapter. The use of the *MakeFeatureLayer* and *MakeTableView* as well as the *Select Layer by Attribute* and *Select Layer by Location* tools will be discussed. The demos and exercise provides an opportunity for the reader to see and develop code that uses queries, feature layers, table views, and the use of optional parameters.

Building the Query Syntax

As mentioned above and shown in Demo 5a, a query expression (sometimes referred to as a "where clause") is often necessary to limit records (such as has been seen in the Make Feature Layer routine **Data Management—Layers and TableViews—Make Feature Layer**). Before discussing the selection routines found in ArcGIS it is necessary to spend some time discussing how to build query syntax using different kinds of data types, since the syntax for queries can vary depending the data type. Building and troubleshooting query syntax is often a challenging and sometimes time consuming practice when developing Python scripts for ArcGIS. Refer to the ArcGIS Help topic **Mapping and Visualization—Working with Layers—Interacting with layer contents—Query expressions in ArcGIS**. In addition the **SQL reference for query expressions using ArcGIS** can be found as links in several locations including **Building a query expression**. These references provide some additional information about building query expressions. See specific tool help with some examples of tools using query syntax. This section bridges many of the concepts found in the ArcGIS help with Python concepts when using queries in the Make Feature Layer, Make Table View, and Select tools. Queries will be discussed again in Chapter 6 which introduces the concept of cursors to search, insert, and update records in a feature class or table.

The reader can refer to the **Demo5a.py** script and the **Sacramento_Streets.shp** file in **\PythonPrimer\Chapter05\Data** folder and can also make some of the changes to the Python script and run it to better understand the kinds of issues that can occur when query syntax is not written correctly. Make sure to copy the script to a new file before making changes. **Demo5a.py** will also be referred to in the next section discussing the Make Feature Layer tool.

Building query syntax may be slightly different, depending on the data type of the source data. Most often, ArcGIS users use shapefiles, personal or file geodatabases, or SDE (Spatial Database Engine) feature classes or tables.

Field Name Syntax

When code developers are writing scripts using file-based data such as shapefiles, file geodatabases, SDE geodatabases, or even ArcIMS feature or image service layers, field names will be enclosed in double quotes.

For example, as used in Demo5a below, the Sacramento streets data is a shapefile and the query used in the Make Feature Layer tool uses double quotes for the field "CLASS." Hence the field name used in a Python query is:

```
query = '"CLASS"...
```

For personal geodatabases, field names should be enclosed in square brackets.

```
query = '[CLASS] = ...
```

If a raster dataset exists within a personal geodatabase, then the field name is enclosed in double quotes similar to the file-based syntax above.

Developing and Processing Strings in Query Expressions

Strings for field names and values are case sensitive, so the code developer needs to make sure of the case for each attribute and value. This is another area that often challenges the code developer.

NOTE: Standardizing field names, values, and case is good practice when designing a geodatabase.

String values are bounded by single quotes, hence:

```
"CLASS" = 'H' or [CLASS] = 'H'
```

The above is typical syntax when using the Query Builder in ArcMap (for example in the Select By Attribute function). Often new Python code developers incorrectly believe they can create and check the query in a Query Builder dialog box and then copy/paste this syntax into a developing script.

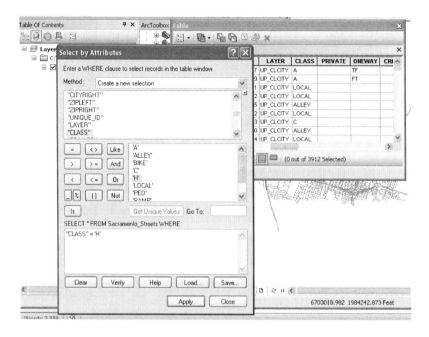

If the above syntax was copied from the ArcGIS query builder and pasted directly into a Python script, the following syntax would appear.

```
        arcpy.Delete_management(feat_layer)

# create a feature layer using a query

query = "CLASS" = 'H'


#query = '"CLASS" = \'H\''
arcpy.MakeFeatureLayer_management(feat_class, feat_layer, query)
```

The query variable will show the word "CLASS" and the character 'H' in green and the equal sign will be shown in black (the reader may need to try this out using an empty Python script to see this effect). Python treats the "CLASS" and 'H' as strings, whereas the equal sign between "CLASS" and 'H' is not part of the string that is supposed to be set to the query variable. If the script was checked with the above syntax, the following error would appear.

In this case Python does not know how to assign the variable query because there are two different equal signs.

Since Python uses single quotes to define strings, even wrapping the right side of the equal sign with single quotes will still not produce the correct syntax.

```
# create a feature layer using a query
query = '"CLASS" = 'H''

#query = '"CLASS" = \'H\''
arcpy.MakeFeatureLayer_management(feat_class, feat_layer, query)
```

Notice in this case that the 'H' is black. Python will treat the green characters as the string and will not know what to do with the "H" character or the two ending single quotes. Note: The reader may need to type the syntax into Python IDLE to see the color effects.

When attempting to check the script with the above syntax for the query, an error message is produced and the query line is highlighted with the problem.

A Python Primer for ArcGIS®

```
if arcpy.Exists(feat_layer):
    arcpy.Delete_management(feat_layer)

# create a feature layer using a query
query = '"CLASS" = H'
arcpy.MakeFeatureLayer_management(feat_class, fea

# print the number of records in the feature layer
result = arcpy.GetCount_management(feat_layer)
print "Number of features in the feature layer " .
```

The solution to fix the syntax is to use the escape character ("\") to properly escape the single quotes in the query string (right side of the query variable).

Without properly using the escape character ("\"), the Python syntax will be erroneous. The challenge, now, becomes where in the query string does the escape character need to be used.

The developer might think that an escape character can be used just before the single quote before the 'H'. Making this change and checking the module also produces an error. Again, Python knows how to interpret the right hand of the equal up to the "H" character, but thinks there is an extra single quote and does not know how to interpret it.

```
if arcpy.Exists(feat_layer):
    arcpy.Delete_management(feat_layer)

# create a feature layer using a query
query = '"CLASS" = \'H''
arcpy.MakeFeatureLayer_management(feat_class, feat_layer, q

# print the number of records in the feature layer with que
```

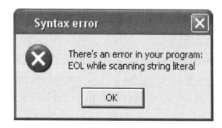

Syntax error

There's an error in your program:
EOL while scanning string literal

OK

To make the Python syntax correct so that ArcGIS understands it, another escape character is required after the ending single quote just after the "H". The following shows the correct Python syntax that will be properly processed by Python and ArcGIS.

```
print "Number of features in the feature layer " + feat_layer +

if arcpy.Exists(feat_layer):
    arcpy.Delete_management(feat_layer)

# create a feature layer using a query
query = '"CLASS" = \'H\''
arcpy.MakeFeatureLayer_management(feat_class, feat_layer, query)

# print the number of records in the feature layer with query
```

The Python script for the query variable line literally does the following:

    a. The single quote starts the query string.
    b. The double quote bounds the attribute name CLASS.
    c. An equal sign is added so that the value for CLASS can be assigned
    d. An escape character "escapes" the single quote telling Python that the single quote is literally translated as a single quote.
    e. The specific value for CLASS (in this case an "H") is assigned.
    f. Another escape character with a single quote is used to indicate that a single quote is required.
    g. A single quote ends the query string.

Common Operators in Queries

As shown above the equal sign is commonly used to query a specific value or values from a list of possible values found in an attribute table. In addition, the following string operators are used when developing query strings for Python and ArcGIS.

LIKE – often used with wild card characters to perform partial string character searches. For example "Street_Name" LIKE 'M%' will return all streets names beginning with the letter "M".

<> – indicates not equal to. This will return all records except the value on the right side of the <>.

>, <, >=, and <= can also be used with character strings. For example, "Street_Name" >= 'M' will return all street names that begin with the letter M through Z.

Wildcard Characters

Wildcard characters are also related to the data type of the data being used.

For file-based data, file geodatabases, and SDE data layers, the following can be used:

% - represents any number of characters. See the example above under the LIKE common operator

_ - (underscore) a wildcard that represents a single character. This value can be placed anywhere in a character string

For personal geodatabase, the equivalent characters are:

* – for any number of characters

? – for a single character

NULL Values

In ArcGIS and databases in general NULL values carry special meaning. NULL represents "No Data" or empty values. NULL values are not zero. NULL values can be used both in vector, tabular, or image data. To use NULL values effectively in Python and ArcGIS, one needs to know how to use them when querying data. When building queries, it might be helpful to query data that is set to NULL or represents NULL (or not). The following shows how NULL can be used to develop queries.

"Street_Name" IS NULL – will query data where the "Street_Name" attribute does not contain a value

"Street_Name" IS NOT NULL – will query data and return records that do contain a value for the "Street_Name" attribute.

In ArcGIS and Python, single quotes are not needed to "bound" the word NULL.

The Python syntax would be:

```
query = '"Street_Name" IS NOT NULL'
```

Numerical Expressions in Queries

Expressions using arithmetic notation can be used to query numerical values. For example,

"City_Pop" > 10000

can be used to query a City_Pop attribute whose values are greater than 10000, where City_Pop is a numerical type attribute field that holds number values.

Standard numerical comparison operators such as =, >, <, >=, <=, <> and the keyword BETWEEN can be used in query strings. In Python the syntax will appear like this:

```
query = '"City_Pop" > 10000'
```

Using Calculations in Queries

The use of numbers in expressions can be expanded by the use of arithmetic operators such as +, -, *, /, and the use of parentheses to group and provide precedence to numerical operations.

For example,

"City_Pop" / "Area" <= 100

Or

"FID" < ("UNIQUE_ID" / 100) + 5

In Python a query like this might look like:

```
query = '"FID" < ("UNIQUE_ID" / 100) + 5'
```

Combining Expressions

Up to this point the examples have shown simple query expressions. In many cases, there may be multiple conditions that a query must meet. In these cases the code developer can use key words to combine expressions.

AND – Both query conditions must be met for the query result to be true to select or return records

OR – At least one condition must be met for the entire query to be true to select or return records

Combining multiple query expressions into a compound query is another area that a code developer can spend considerable time working through to create the right syntax and return the correct set of records or features.

A good approach for making progress on more complex queries is to start with one condition at a time, validate the syntax and returned set of queried data, and then build the next portion of the query. Compound query strings can be very long, may include a combination of character strings, numbers, NULL values, variable names, etc. providing plenty of opportunity for errors.

Recommendations

Building complex queries is one place that spending time within ArcMap to manually work through the query steps can benefit the development of coding logic and syntax. If the code developer can see what actual attributes and values are needed as well as to see the results of a query in selected records that can be interactively viewed and examined while in ArcMap, he/she can gain insight to the specific ArcGIS and Python syntax (such as the data type, wildcard and escape characters, and keywords) that will be needed to write the queries correctly. Another recommendation is the use of print statements within the Python script to report back the results of the number of records or features which can be compared to the results from manual processes developed using ArcMap.

Other Query Syntax

The following topics are not specifically covered in *A Python Primer for ArcGIS*, however, readers are urged to review the ArcGIS Help topic *SQL reference for query expressions used in ArcGIS*. Some of these are dependent on the type of database being used and are considered more advanced procedures.

Dates – e.g. Date, Date and Time, Time, Day, Month, Year parts of date, etc. Specifically see the section for Date and Time.

Numerical Functions – e.g. sin, cosine, round

String Functions – e.g. finding the left most, right most, or middle part of a string value

Some of these may be discussed in other chapters, but not in a comprehensive way.

Feature Layers and Table Views

Feature classes and tables represent actual files that can be found on disk or within a geodatabase. These can be in the form of shapefiles in a folder structure or special tables in geodatabases (personal, file, or ArcSDE) that contain the "SHAPE" field. Tables can be in the form of a standalone table (such as a dBase file) or a table within a geodatabase that does NOT contain the "SHAPE" field. Alternatively, feature layers and table views are "in memory" occurrences of feature classes and tables. No "new" data is created with feature layers or table views, however, subsequent geoprocessing tools can be used to create new feature classes (**Data Management—Features—Copy Features**) and tables (**Data Management—Tables—Copy Rows**) from them. These will be discussed later. Users of ArcGIS often work with feature layers and table views when adding and working with geographic features or tables when using ArcMap. Feature layers and table views are actually shown in the Table of Contents of ArcMap. The user may not be aware that they are using feature layers or table views. Feature layers and table views are only present during an active ArcMap or ArcCatalog session or model or during the action of running a Python script. Once ArcMap, ArcCatalog, or the model is closed or the Python script finishes processing, the feature layer or table view is removed from memory. In an active Python script, a code developer can use the `Delete_management` function (found under **Data Management—General** in ArcToolbox) to remove access to the feature layer (and its respective memory on the computer) if needed. A code developer may find this useful when the script iterates through a number of feature classes in a folder or geodatabase.

The following illustration shows the Table of Contents of ArcMap showing "feature layers" in the table of contents.

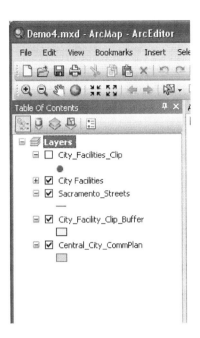

Likewise, table views can be seen by choosing the "List by Source" button from the Table of Contents.

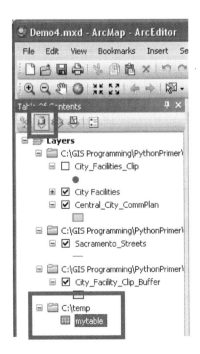

When interacting directly with ArcGIS a user simply adds a feature class or table to the table of contents. Once either kind of data is added, the feature layer or table view is automatically created and the respective layers and tables can be used in ArcToolbox tools and ModelBuilder. When developing standalone Python scripts such as those discussed in *A Python Primer for ArcGIS*, the feature layer or table view must be created as a separate step before feature layers or table views can be used in other geoprocesses. From a Python script development point of view, the Make Feature Layer and the Make Table View geoprocessing tools are used before writing syntax for the selection tools mentioned above. See the ArcGIS Help for each of the respective tools.

Make a Feature Layer

The Make Feature Layer tool generates the "in-memory" representation of a feature class which provides the structures to geoprocessing scripts so that the information in feature classes can be accessed (such as queries, creating selection sets, cursors, and joining tables). The Make Feature Layer tool is shown below. The only required parameters are the feature class name and the name of the layer, which is just a name (character string in Python). The other parameters, such as the query (expression) string, are optional.

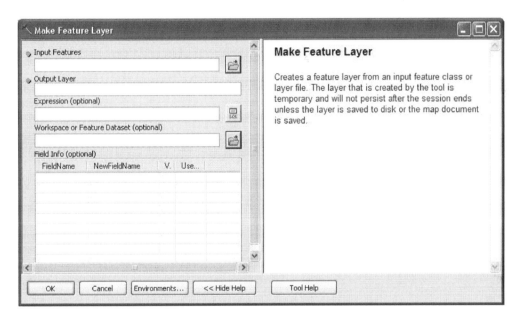

For many GIS operations, only the input feature class and output feature layer name will be required. The expression allows one to limit the number of features available for other operations. The field names parameter can be used to hide existing fields or change field names in the resulting feature layer.

Once the Make Feature Layers tool is used to create a "feature layer," the Select Layer by Attribute or Select Layer by Location tool (as well as other tools) can be used.

Make a Table View

In a similar manner to creating a feature layer, a table view (or in-memory representation) of a table can be created using the Make Table View tool. The following shows the Make Table View dialog box. Its functions are essentially the same as Make Feature Layer, but it operates on tables rather than feature classes.

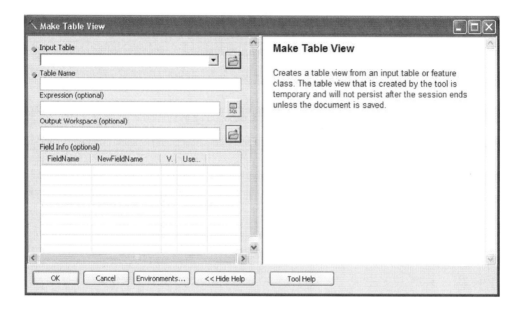

Other operations such as joining feature classes to other feature classes or tables will be discussed in a Chapter 6. Chapter 5 focuses on the fundamental concepts required to create and use feature layers and table views, as well as selecting features or rows from them, respectively.

Refer to the end of the chapter for **Demo 5a**. The reader is encouraged to follow the steps and practice actually writing and running the demo code to understand the basic construction of an attribute query and the Make Feature Layer and other Python concepts. The fully functioning Python script is available under the **\Chapter05** folder.

Selecting Data

The main reason to select data is so that other geoprocessing operations can occur. For example, the user may need to calculate a value for the selected records or export the selected data to a new feature class or table. Selecting data (either by attribute or by location) uses the structures mentioned above.

Select Data by Attribute

To select data by attribute the user implements an attribute query (that can be simple or complex) on the attribute table of the feature class or table with the use the respective feature layer or table view. As shown above, a feature layer or table view can also accept a query as one of the parameters. Feature layers or table views can be used with other geoprocessing functions such as calculating fields or exporting features or records to new data files or tables. However, the select routines have additional functionality that expands the capability of selecting records. Whereas the Make Feature Layer or Make Table View routines only have a query parameter to limit the number of features or records, the Select Layer by Attribute has a number of methods to select data. Shown below is the `SelectLayerByAttribute` routine. The required parameter is an input feature layer or table view. The optional parameters are the selection method and "where clause" parameters. Although the selection method is an optional parameter, if it is not specifically provided to the `SelectLayerByAttribute` routine, the "NEW_SELECTION" method will be used. The "where clause" parameter will often use a variable that has the defined query for the selection.

```
arcpy.SelectLayerByAttribute_management
(<input_feature_layer_or_table_view>, {selection method},
{where clause})
```

The following summarize the selection methods.

>NEW_SELECTION – create a new selection; default selection method
>ADD_TO_SELECTION – add selected records to existing selected records

REMOVE_FROM_SELECTION – remove records from existing selection
SUBSET_SELECTION – combine selected records that are common to the existing selection
SWITCH_SELECTION – select records that are not already selected and unselect the existing selected records
CLEAR_SELECTION – clear all existing selected records

The Make Feature Layer (or Make Table View) routine does not include this functionality and would be very cumbersome to make changes to queried records. In addition, when a Make Feature Layer (or Make Table View) is used, only the queried records are available for further use. One can think of the Make Feature Layer (or Make Table View) routine with a query similar to a "Definition Query" from the Layer Properties—Definition Query tab.

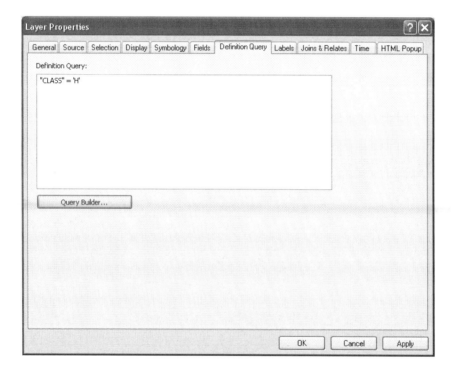

The result of the definition query is shown in the map and layer's attribute table.

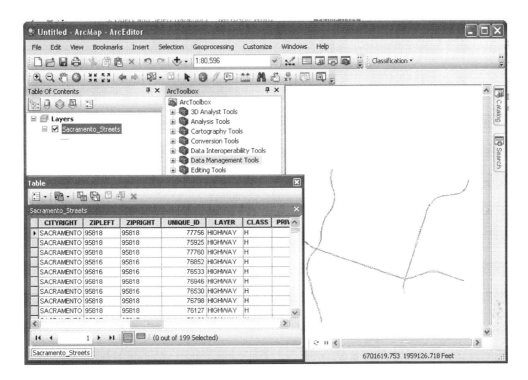

Notice above that only the highways are showing in the map and that all of the records in the attribute table show the "CLASS" attribute with the value of 'H'. Note the total number of records (199). If the Make Feature Layer function was implemented in Python using the query parameters described above and shown in Demo5a, the same number of records would result. Because the feature layer has a query in it, the other records from the Sacramento_Streets feature class are not available for selection, computations, or other geoprocesses unless the feature layer is deleted and a new query is implemented. The Select Layer by Attribute provides a better and more flexible method to easily access all of the data within a feature class or table.

To illustrate the `SelectLayerByAttribute` routine functionality a user can run the **Data Management—Layers and Table Views—Select Layer by Attribute** tool and fill in the following parameters. Note, the Definition Query was removed before using this tool.

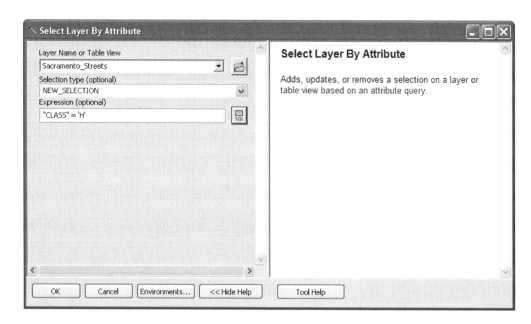

When the user clicks OK, the following map and attribute table are shown.

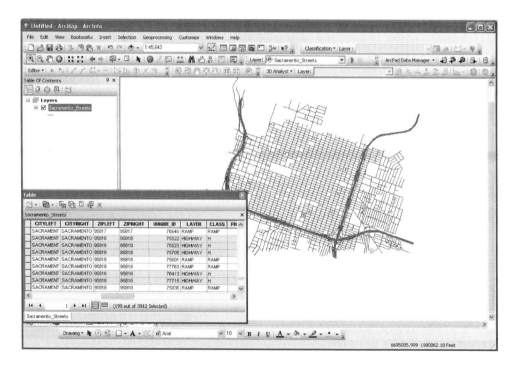

The `SelectLayerByAttribute` actually "selects" (highlights) the features or records (in this case 199 of 3912) that meet the query criteria from the feature layer (or table view). As shown above, all of the data are available to perform other queries and selections. Subsequent "selections" can be performed on the same layer by implementing the SelectLayerByAttribute again and choosing a different selection method and/or changing the query expression.

Programming the *SelectLayerByAttribute* Routine

In Python to implement the SelectLayerByAttribute routine, the code developer simply adds the SelectLayerByAttribute tool to the script after the MakeFeatureLayer (or MakeTableView) tool. Refer to the **Demo5b.py** script.

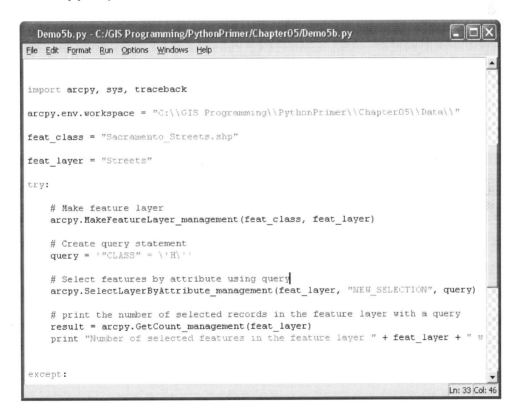

As shown above, the first step is to create a feature layer. Alternatively, the code developer can create a variable to hold the query string. Using a variable to hold the query statement can help make other Python statements easier to read and make changes. The next step is to add the `SelectLayerByAttribute` routine that uses the feature layer as the input (which is the required format for the first parameter), the type of selection (selection method), and the optional query string. Typically, once a selection is made, another geoprocessing operation does something with them. For this simple example, the number of selected records are counted and printed to the Python Shell. A more realistic result is shown below.

Select Data by Location

A different method of selecting records can be done by evaluating the spatial relationship between one geographic layer and another. A number of built-in operations exist to extract or merge geospatial data that has a spatial relationship such as clip, buffer, intersect, union, spatial join among others. One option that these functions do not have is the ability to select records from existing data and perform a subsequent operation on the selected records. For example, the buffer routine cannot select features on a separate feature class. The select features by location, however, can select features based on the spatial coincidence of an existing boundary or by applying a "virtual radius" around features from the same feature class. Both operations can be performed in a single step. Shown below is the general syntax for the `SelectLayerByLocation` routine.

```
SelectLayerByLocation_management (<input_feature_layer>,
{spatial_relationship_type}, {select_features},
{search_distance}, {selection_method})
```

The only required parameter is an input feature layer. Table views are not used in the `SelectLayerByLocation` routine, since it operates using a spatial relationship between two layers. The optional parameters include:

Spatial_relationship_type – the default is INTERSECT, but can be one of many different types of spatial relationships and depends on the feature type of the datasets. See the ArcGIS Help for the `SelectLayerByLocation` routine.

Select_features – the feature layer that will be used to perform the feature selection on the input feature layer.

Search_distance – the distance from features to expand the selection on the input feature layer. This parameter will have a number and a unit type (e.g. 100 METERS).

Selection_method – the type of selection. These choices are the same as the `SelectLayerByAttribute`.

If the optional parameters are not specifically added, the INTERSECT spatial relationship and "NEW_SELECTION" selection method will be used. In addition, a search radius of zero units will be applied.

To get an idea of how the `SelectLayerByLocation` tool works, the user can open the **Data Management—Layers and Table Views—Select Layer by Location** tool and add set the following parameters. The reader can follow this example by using data from the **\PythonPrimer\Chapter05\Data** folder. In this example, the analyst wants to determine how many parcels are within 200 feet of highways.

NOTE: In this example Sacramento_streets is assumed to have only the highways selected which resulted from a `SelectLayerByAttribute` routine (see the previous section).

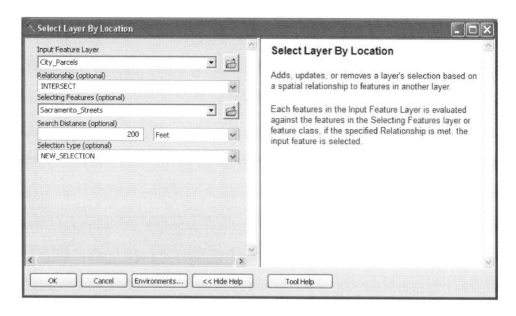

In the Select Layer by Location tool above, the following parameters are set:

Input Feature Layer - City_Parcels layer is the input future layer that will have the selection applied to it.

Relationship - INTERSECT, which tells ArcGIS to select all parcels that are spatially coincident with the "Selecting Features" parameter. INTERSECT is the default. Check the ArcGIS Tool Help for other options.

Selecting Features – the Sacramento_Streets is the feature layer used to actually select features in the input feature layer

Search Distance (and units) – the value used to determine an additional search radius around the selecting features layer

Selection Type – NEW_SELECTION, the type of selection being applied to the input feature layer. NEW_SELECTION is the default; the other options are the same as the Select Layer by Attribute. See the ArcGIS Tool Help for more details.

After clicking OK, the following map and attribute table for the City_Parcels layer result.

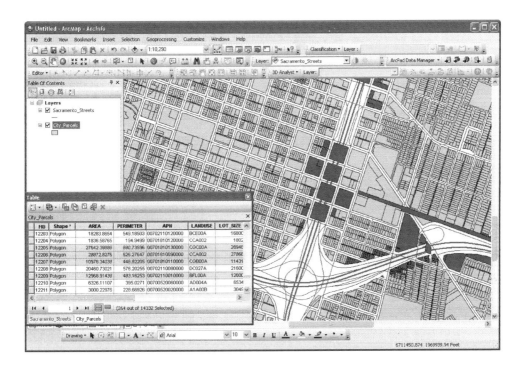

The reader can see that only parcels that are within a distance of 200 feet from the selected highway features are selected (shown in dark color). The accompanying City_Parcels attribute table shows some of the selected parcels.

Programming the *SelectLayerByLocation* Routine

In Python, the reader can expand upon the geoprocessing tasks presented in **Demo5b.py** by adding a few lines of code to implement the `SelectLayerByLocation` routine. Since the `SelectLayerByLocation` routine is selecting data from a separate feature layer, two new variables are created, one for the parcel feature class and one for the feature layer. An additional variable is provided that holds the search distance. These are defined toward the top of the Python script to keep the script organized.

```
poly_fc = "City_Parcels.shp"
```

```
            poly_layer = "City Parcels"
            search_distance = '200 FEET'
```

Because the `SelectLayerByLocation` routine requires feature layers, a new `MakeFeatureLayer` line is added to create a feature layer from the City_Parcels.shp feature class. The code below shows the **Demo5b.py** script with these additions.

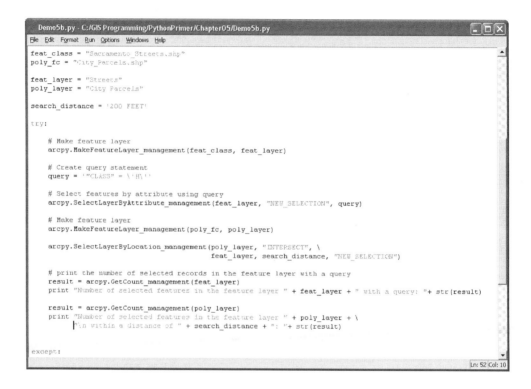

The **Demo5b.py** script above shows the functioning code using a custom query, the `MakeFeatureLayer` routine to create feature layers for the required feature classes as well as the `SelectLayerByAttribute` and `SelectLayerByLocation` with same parameters as described above in this chapter. The script also reports back the number of selected features for each feature class. In practice, other geoprocessing functions are implemented. See the section below for creating new datasets.

In addition to the script showing the selection methods, a couple of new Python syntax elements are used. Notice in the `SelectLayerByLocation_management` line that a single backslash ("\")

is used after the spatial relationship type "INTERSECT". The single backslash allows for Python syntax for a single operation to continue on a separate line. The code developer may find this useful to prevent from having to write out long single lines of code. The backslash is also used in the last print statement. Also, in the last print statement, is the use of the "\n". This character will cause a `print` statement to continue on the next line of the Python Shell. This can help prevent long message from being written out on a single line.

Counting the Number of Records

The reader has already seen the use of the Get Count tool (**Data Management—Table—Get Count**) that is useful for determining how many features or records there are in a data set or selection set. As shown in several of the examples above the result of Get Count can be assigned to a variable that was used in print statements. This can be useful to help code developers to track and check results from selection queries. The value from Get Count can also be used in subsequent processes such as setting a bound on a `for` loop or a number of iterations for a `while` loop. The reader will see the use of Get Count throughout this book and in the example scripts.

Creating a New Dataset

As mentioned above one possible subsequent operation after a select layer operation is to write the selected records to a new feature class or table. If a feature class is written, then the Copy Features tool (**Data Management—Features—Copy Features**) is used to write out both the geographic features and accompanying attribute table. If only the attribute table or a standalone table is desired, then the table operation Copy Rows (**Data Management—Tables—Copy Rows**) is used. Each of them is demonstrated in the **Demo5b.py** script. See the ArcGIS Tool Help for the Copy Features or Copy Rows tools as needed.

```
try:
    # Make feature layer
    arcpy.MakeFeatureLayer_management(feat_class, feat_layer)

    # Create query statement
    query = '"CLASS" = \'H\''

    # Select features by attribute using query
    arcpy.SelectLayerByAttribute_management(feat_layer, "NEW_SELECTION", query)

    # Make feature layer
    arcpy.MakeFeatureLayer_management(poly_fc, poly_layer)

    arcpy.SelectLayerByLocation_management(poly_layer, "INTERSECT", \
                                           feat_layer, search_distance, "NEW_SELECTION")

    # print the number of selected records in the feature layer with a query
    result = arcpy.GetCount_management(feat_layer)
    print "Number of selected features in the feature layer " + feat_layer + " with a query: "+ str(result)

    result = arcpy.GetCount_management(poly_layer)
    print "Number of selected features in the feature layer " + poly_layer + \
          "\n within a distance of " + search_distance + ": "+ str(result)

    # Copy the selected polygon features to a new shapefile
    arcpy.CopyFeatures_management(poly_layer, out_poly_fc)
    print "Copied selected features from " + poly_fc + " to " + out_poly_fc

    # Copy the selected street segment attributes to a dBase table
    arcpy.CopyRows_management(feat_layer, out_street_attributes)
    print "Copied selected attributes from " + feat_class + " to " + out_street_attributes

    print "Completed Script"
except:
```

For a full detailed set of steps for **Demo5b.py** refer to the **Demo5b** section below.

Data Locks

When analysts work with datasets where the data values will be created, updated or changed, or deleted some care is required to make sure that the dataset's integrity is maintained. Typically, an analyst is working in either ArcMap to interact with records or fields of data sets or ArcCatalog to add/or delete fields, their data types, and possibly attribute domains. In either case, when an application accesses the dataset, effectively a data lock is placed on it indicating that another application cannot make changes (to either records, fields, data types, attribute domains, etc). For example, if an analyst has a streets layer open in ArcMap and then tries to access the dataset in ArcCatalog to add a field, the following message will appear.

With file geodatabases, when a feature class, table, raster, or other data set is accessed through an application, a .LOCK (lock file) will be added. The .LOCK file can be seen through Windows Explorer.

Essentially, data locking mechanisms are put in place as a measure of protection so that analysts do not inadvertently change their data and to provide a message back to the user letting them know the data is being used by another application or user. When one of the applications (such as ArcMap) is closed, the lock is released and changes can be made to the respective data set. Only one application can access the data if changes are to be made.

From a programming point of view, especially when developing code, similar kinds of measures are put into place, but are not always apparent to the code developer. In addition, since a code developer often writes a geoprocess that creates a result, he or she often wants to review the data in ArcCatalog or ArcMap to make sure it was created properly. Running Python to process data is an "application" that uses geospatial data and thus, data locks can be created. Even if Python is the only application accessing geospatial data, data locks can persist with the following conditions:

1. When a script successfully completes processing, but Python (IDLE and the shell) is not closed.

2. The Python script fails because of syntax or processing logic problems and the analyst attempts to re-run the script again without closing Python (IDLE and shell), even if changes are made to the script and it is re-saved.
3. The script creates/updates a data set and the analyst reviews it in ArcMap or ArcCatalog, Python is closed, but ArcMap or ArcCatalog remain open, and the analyst attempts to re-run a script that creates/updates the same data set that is open in ArcMap or ArcCatalog.

In order to prevent data locks that prevent code developers from reviewing or reprocessing data, the following procedure can be followed:

1. Do not have ArcMap or ArcCatalog open (any occurrence) when processing data using Python IDLE (or other script editor to run the script outside of ArcMap or ArcCatalog).
2. After Python successfully (or unsuccessfully) processes a data file, close Python IDLE and the Python Shell (all occurrences) before reviewing at data derived from Python processes in either ArcMap or ArcCatalog. NOTE: If the code developer noticed error messages in the Python Shell, it is recommended to do a screen shot of the error or copy and paste the error message into a text editor before closing Python IDLE and the Python Shell.
3. After Python successfully (or unsuccessfully) processes a data file and the code developer makes changes to the script (or re-runs the script without any changes) that re-creates or updates the same data set (even if the developer does not review it in ArcMap or ArcCatalog), close Python IDLE and the Python Shell (all occurrences) before re-running the script.

One might think that using the `Exists`/`Delete` functions or the `overwriteOutput` environmental option (see the ArcGIS Help for the `overwriteOutput` routine) can be used in scripts and geoprocesses that takes care of data locking, but data locking has more to do with more than one application (or multiple occurrences of the same application) accessing the same data. Think of trying to edit the same Word or Excel document in more than one occurrence of Word or Excel. The application often reports to the user that the document is open in another location. With GIS data, ArcGIS, and Python, it is the same kind of issue.

The following error message is produced as a result of the data set being opened in ArcMap while attempting to add a field to it using a Python script.

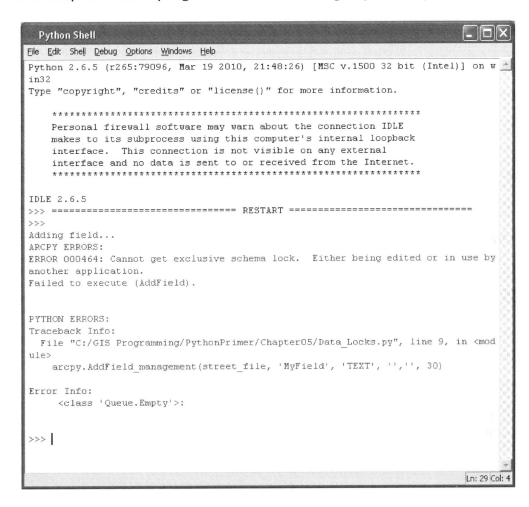

In this case, the code developer would need to exit the ArcMap document that is using the dataset before re-running the script.

Data locking is mentioned in this chapter so that the reader is aware of the concept and how it can impact the development and processing of code as well as attempting to review data derived from a Python script so that troubleshooting can take place. Data locking will occur more often when specific records are being updated and created (such as with the use of cursors in Chapter 6).

Summary

Chapter 5 discussed and reviewed the primary mechanisms for create and using queries in the `SelectLayerByAttribute` and `SelectLayerByLocation` routines, the two most common methods for querying and selecting data. Understanding the requirements of using the `MakeFeatureLayer` and `MakeTableView` routines in addition to constructing the proper syntax for queries is a key to using the selection methods. In many instances, the code developer will be required to construct a query to limit a selection, a set of records, a number of elements that will be used in subsequent geoprocessing steps. Properly building and using queries in addition to the use of the `MakeFeatureLayer` and `MakeTableView` are fundamental to many of the geoprocessing workflows to develop standalone Python scripts.

In addition, this chapter also introduced counting records and creating new datasets, which happens to be a common "next step" after selecting a subset of data. Data locking was also mentioned and will be a key element to pay attention to when the code developer tests and troubleshoots problems. Being able to recognize that ArcGIS and the Python Shell and IDLE can cause locks on data and will require proper suspension when attempting to solve problems with a script, data, and processing logic.

Demos Chapter 5

The Chapter 5 demos are divided into two sections. **Demo 5a** focuses on using the `MakeFeatureLayer` routine and showing how to use the routine with and without the query parameter. **Demo 5b** focuses on extending these concepts to selecting features from a dataset. A primary objective of the reader is to gain an understanding of how to properly construct attribute queries using the proper Python and ArcGIS syntax so that the query can be used in the `SelectLayerbyAttribute` and the `SelectLayerByLocation`, the two commonly used methods of selecting data records.

Demo 5a: Create a Feature Layer with and without a Query

This demonstration takes a look at how the Make Feature Layer Tool is used to create a feature layer from a feature class with and without a query. Refer to **Demo 5a.py** for the actual Python script. See the respective help for the various tools described below for additional information.

The concepts illustrated in this demo are:

ArcGIS Concepts

Workspace
Building a Query
Create and Use a Feature Layer
Counting Records
Building a Query
`Exists` Method
`Delete` Method

Python Concepts

Variable definitions
`If` statement
Escape Character for a Single Quote (\')
Use of `try:` and `except:`
Use of `import sys` and `traceback` modules

This demo uses the **Sacramento_Streets.shp** file that can be found in **\PythonPrimer\Chapter05\Data** folder. Open ArcMap or ArcCatalog to see the various attributes and values of the **Sacramento_Streets.shp** file if needed. A **Demo5.mxd** has been provided.

1. Start out by importing the `arcpy`, `sys`, and `traceback` modules. Add some commentary text to briefly describe the script.

2. Add in the `try` and `except` blocks. (HINT: Use the exception code found in **\Chapter01**).
3. Add a line to set the workspace and some variables to hold the values of the feature class and feature layer. Note that the `feat_layer` variable below is just a string of characters and can represent any name the developer chooses.

NOTE: The path to the \PythonPrimer\ChapterXX\Data may need to be modified, depending on where the reader copied the data for the book.

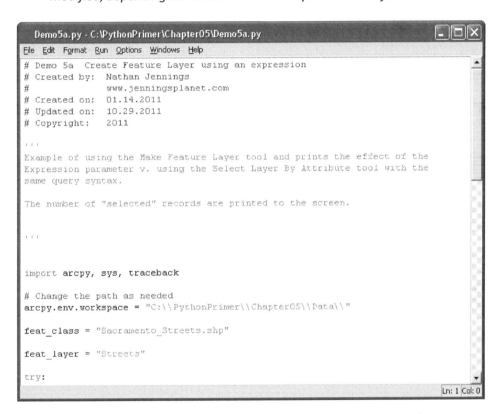

Before creating a layer, it might be good to get an idea of how many features are actually in the feature class. This can be done by using the `GetCount` routine from the **Data Management—Table—Get Count** tool.

4. Add the lines of code to report back the number of features (records) in the **Sacramento_Streets** feature class.

```
Demo5a.py - C:\PythonPrimer\Chapter05\Demo5a.py
File  Edit  Format  Run  Options  Windows  Help
# Copyright:    2011

'''
Example of using the Make Feature Layer tool and prints the effect of the
Expression parameter v. using the Select Layer By Attribute tool with the
same query syntax.

The number of "selected" records are printed to the screen.


'''

import arcpy, sys, traceback

# Change the path as needed
arcpy.env.workspace = "C:\\PythonPrimer\\Chapter05\\Data\\"

feat_class = "Sacramento_Streets.shp"

feat_layer = "Streets"

try:

    # print the number of records in the feature class
    result = arcpy.GetCount_management(feat_class)
    print "Number of features in the feature class " + feat_class + " : " + str(result)
```

The developer will find that `GetCount` returns the number of records in the feature class. (`GetCount` can also be used with a feature layer, table, or table view). The above code shows this value being assigned to the `result` variable. Since this value is assigned to a variable, the actual number of records can be used in other parts of a script. In this case the `result` variable is used in a print string to print the number of records back to the Python Shell. Notice also, that `result` is a number. In order to "concatenate" a number with a string, it must be converted (or cast) as a string, hence the use of `str(result)`. In addition, notice how the `feat_class` variable is used in `GetCount` (i.e. to get the number of records in the feature class (in this case) and in the print statement. The name of the feature class will be printed in the print statement as a result of using the `feat_class` variable.

5. Create a Feature Layer without a Query

After the number of records in the feature class is reported, the line creates the feature layer by using the `MakeFeatureLayer` routine. In addition, similar lines of code can be written to report back the number of records in the feature layer.

```
Demo5a.py - C:\PythonPrimer\Chapter05\Demo5a.py
File Edit Format Run Options Windows Help

import arcpy, sys, traceback

# Change the path as needed
arcpy.env.workspace = "C:\\PythonPrimer\\Chapter05\\Data\\"

feat_class = "Sacramento_Streets.shp"

feat_layer = "Streets"

try:

    # print the number of records in the feature class
    result = arcpy.GetCount_management(feat_class)
    print "Number of features in the feature class " + feat_class + " : " + str(result)

    arcpy.MakeFeatureLayer_management(feat_class, feat_layer)

    # print the number of records in the feature layer without a query
    result = arcpy.GetCount_management(feat_layer)
    print "Number of features in the feature layer " + feat_layer + " without a query: "+ str(result)

Ln: 18 Col: 0
```

Notice that the `MakeFeatureLayer` routine uses the feature class as the input and the name of the feature layer. These two parameters are the only required parameters to create a feature layer from a feature class. The optional parameters can be ignored for this routine (such as a query expression). *(Additional comments will be made later to describe using a "place holder" value for optional parameters to help maintain the correct order, number, and type of parameters in a geoprocessing function).*

For the `GetCount` and `print` statements notice that the `feat_layer` variable is used instead of the `feat_class` variable. Using the `feat_layer` variable allows `GetCount` to obtain the number of records from the feature layer (rather than the feature class) and subsequently print the number of records to the screen (or use the number for other geoprocesses).

6. Create a Feature Layer using a Query

The next set of lines will set up the ability to use a query in `MakeFeatureLayer` routine. Since the next feature layer will create another feature layer for the same feature class, the developer can simply "re-use" the variable already created and used in the previous steps. The previous

`MakeFeatureLayer` routine already used the name "Streets." Attempting to "re-use" the same name will cause the program to result in an error indicating the feature layer already exists. An option would be to set another unique variable to a unique feature name, however, this can be avoided by using an `if` statement to check if the feature layer name exists using the `Exists` routine; if it does, then the feature layer name can be deleted by using the `Delete` routine. Using an `if` statement and checking to see if a data sets exists is common practice in writing Python scripts and can help in re-using variables and overwriting various datasets (feature classes, feature layers, tables, table views, work spaces, images, etc). Notice how indenting is used to make sure the `if` statement is processed correctly.

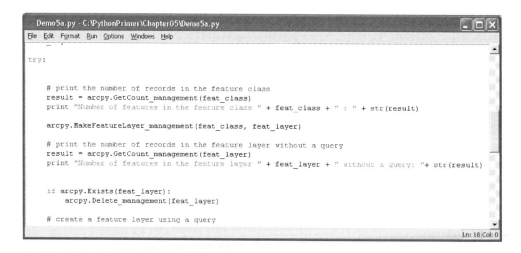

After the `if` block is written, the query expression string for the `MakeFeatureLayer` can be written. Shown in the code below is a variable that stores the syntax for the query.

```
query = '"CLASS" = \'H\''
```

Notice that the query variable is used as the query parameter in the `MakeFeatureLayer` routine which results in a more concise script. See below.

```
Demo5a.py - C:\PythonPrimer\Chapter05\Demo5a.py
File Edit Format Run Options Windows Help
arcpy.env.workspace = "C:\\PythonPrimer\\Chapter05\\Data\\"

feat_class = "Sacramento_Streets.shp"

feat_layer = "Streets"

try:

    # print the number of records in the feature class
    result = arcpy.GetCount_management(feat_class)
    print "Number of features in the feature class " + feat_class + " : " + str(result)

    arcpy.MakeFeatureLayer_management(feat_class, feat_layer)

    # print the number of records in the feature layer without a query
    result = arcpy.GetCount_management(feat_layer)
    print "Number of features in the feature layer " + feat_layer + " without a query: "+ str(result)

    if arcpy.Exists(feat_layer):
        arcpy.Delete_management(feat_layer)

    # create a feature layer using a query

    query = '"CLASS" = \'H\''


    #query = '"CLASS" = \'H\''
    arcpy.MakeFeatureLayer_management(feat_class, feat_layer, query)
```

The `MakeFeatureLayer` routine above shows that only street features that are of "class H" (Highways) will result in the feature layer. "CLASS" is an attribute where one of the values is 'H'. Open ArcMap or ArcCatalog to see the various attributes and values of the **Sacramento_Streets.shp** file if needed. The same `GetCount` and print statement can be used to print out the number of records in the feature layer with the query. Notice that the query and subsequent lines are not indented like the `Delete` routine. These lines are NOT part of the `if` block that check to see if the feature layer exists.

```
# print the number of records in the feature class
result = arcpy.GetCount_management(feat_class)
print "Number of features in the feature class " + feat_class + " : " + str(result)

arcpy.MakeFeatureLayer_management(feat_class, feat_layer)

# print the number of records in the feature layer without a query
result = arcpy.GetCount_management(feat_layer)
print "Number of features in the feature layer " + feat_layer + " without a query: "+ str(result)

if arcpy.Exists(feat_layer):
    arcpy.Delete_management(feat_layer)

# create a feature layer using a query

query = '"CLASS" = \'H\''


#query = '"CLASS" = \'H\''
arcpy.MakeFeatureLayer_management(feat_class, feat_layer, query)

# print the number of records in the feature layer with query
result = arcpy.GetCount_management(feat_layer)
print "Number of features in the feature layer " + feat_layer + " with a query: "+ str(result)



except:
```

7. Save, Check, and Run the script

If the reader has been writing the code for this demo, save the script and run the Check Module. Fix any syntax issues as needed. Once all of the errors are fixed, run the script from within IDLE. The print statements should print to the Python Shell. If needed consult the **Demo5a.py** script to help fix problems. The **Demo5 script results.doc** document can be compared with the reader's results.

Demo 5b: Select Features by Attribute

This demo expands the concepts of **Demo 5a** and implements the Make Feature Layer geoprocessing tool and uses it to perform both an attribute selection and a select by location operation. This demo illustrates the following concepts:

ArcGIS Concepts

Building a Query
Creating and Using a Feature Layer
Select data by Attribute
Select data by Location
Counting Records
Creating a Feature Class
Creating a Standalone Table

Python Concepts

If statement
Escape Character for a Single Quote (\')
Line continuation character ("\")
New line character ("\n")
Use of `try` and `except`
Use of `import sys` and `traceback` modules

This demo uses the **Sacramento_Streets.shp** and **City_Parcels.shp** files that can be found in **\PythonPrimer\Chapter05\Data** folder. Open ArcMap or ArcCatalog to see the various attributes and values of the shapefiles if needed. Refer to the Chapter 5 text above for additional discussion and commentary. Two files are created as output in this demo, **out_parcels.shp** (contains the selected features and attribute table from the Select Layer by Location routine) and **out_street_attributes.dbf** (contains the selected attribute records from the Select Layer by Attribute routine).

1. Start the script by importing the arcpy, sys, and traceback modules. Add some commentary to the beginning of the script.
2. Add the `try:` statement and copy the exception code to the `except:` block from the **Exception.py** script found in **\Chapter01**.
3. Add the line for the workspace to the data path **\PythonPrimer\Chapter05\Data**.

   NOTE: The path to the \PythonPrimer\ChapterXX\Data may need to be modified, depending on where the reader copied the data for the book.

4. It is also a good idea to put in some pseudo-code to outline some of the tasks that are expected in the script.

Remember to follow proper indentation and case.

The **Demo 5b** script should look similar to this up to this point.

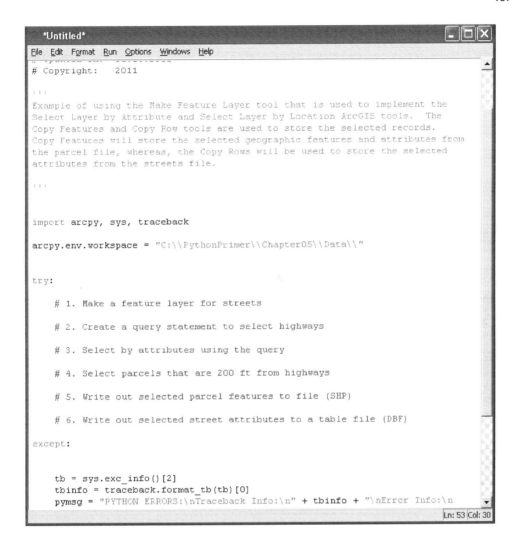

To get started with writing the different geoprocesses it is a good idea to research and review the specific geoprocessing tool help. Refer to the ArcGIS Help as needed.

For the Make Feature Layer, a feature class and a feature layer are required parameters. It is probably a good idea to define some variables for both the feature class and feature layer.

NOTE: When first starting out with writing code, it might be a good idea to write in the "hard coded" values to get an idea of what they might be. Later, when reviewing the script, variables can be created, especially if the same hard coded values are repeated. This is an indicator that a variable might be a good idea.

5. Create two variables above the try: statement.

```
feat_class = "Sacramento_Streets.shp"
feat_layer = "Streets"
```

6. Add the Make Feature Layer tool using the variables as parameters in the routine.

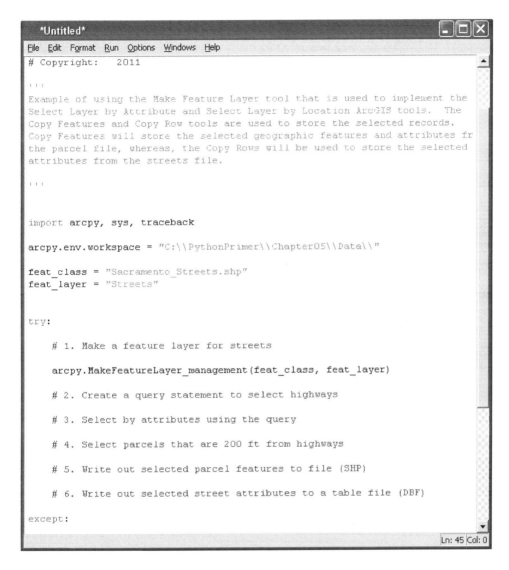

Next, since the Select Layer by Attribute will require the highways to be selected, a variable will be set up to hold the query statement. This is good practice so that the query statement can be isolated from the Select Layer by Attribute routine and keep the code "clean and concise."

7. Create a variable for the query and assign the variable to the proper query syntax.

NOTE: Creating the correct syntax may require the use of ArcMap and the respective data layers (in this case the **Sacramento_Streets.shp** file) to test the query statement logic. Open up ArcMap and add the **Sacramento_Streets.shp** file. Run the Select Layer by Attribute tool or the Select Layer by Attribute option from the Selection menu. The query should use the "CLASS" attribute to select all segments that have a value of 'H'. Hence, the query should look like the following within the tool.

"CLASS" = 'H'

This same syntax will be used to set up the query statement in Python. See the commentary within this chapter that discusses the proper Python syntax for the query expression.

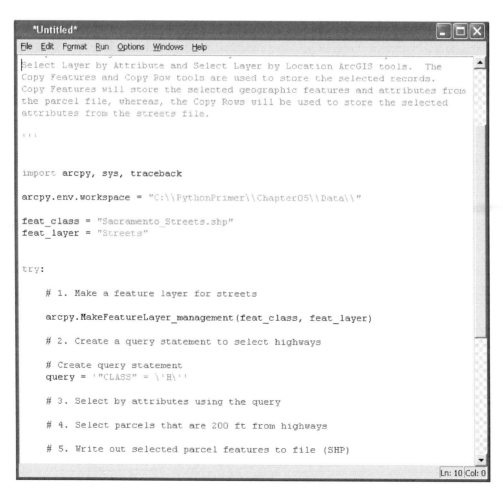

8. Add the Select Layer by Attribute code.

Now that a query statement has been defined, the Select Layer by Attribute routine can be added. Review the tool help as necessary and add the proper parameters. Since this is the first time using this tool, the selection method will be "NEW_SELECTION".

9. Set up parameters for the Select Layer by Location.

Reviewing the ArcGIS Help on Select Layer by Location, the developer finds that another feature layer is required. Before the Select Layer by Location can be implemented, another Make Feature Layer routine is needed.

a. Add a Make Feature Layer line that will create a feature layer from the City_Parcel.shp file.
b. Add in two new variables similar to those above. Put the variable definitions above the try: block.

```
poly_fc = "City_Parcels.shp"
poly_layer = "City Parcels"
```

c. Use the variable definitions in the Make Feature Layer similar to those as the Sacramento_Streets layer. Notice that the "pseudo-code" is slightly modified to include the new step (4a). Updating and making notations to the pseudo-code is good practice to keep in-line code documentation up-to-date. Adding comments with date stamps can indicate when additions and modifications to code were made.

```
*Untitled*
File Edit Format Run Options Windows Help

import arcpy, sys, traceback

arcpy.env.workspace = "C:\\PythonPrimer\\Chapter05\\Data\\"

feat_class = "Sacramento_Streets.shp"
poly_fc = "City_Parcels.shp"

feat_layer = "Streets"
poly_layer = "City Parcels"


try:

    # 1. Make a feature layer for streets

    arcpy.MakeFeatureLayer_management(feat_class, feat_layer)

    # 2. Create a query statement to select highways

    # Create query statement
    query = '"CLASS" = \'H\''

    # 3. Select by attributes using the query

    # Select features by attribute using query
    arcpy.SelectLayerByAttribute_management(feat_layer, "NEW_SELECTION", query)

    # 4. Select parcels that are 200 ft from highways

    # 4a. Make feature layer

    arcpy.MakeFeatureLayer_management(poly_fc, poly_layer)

    # 5. Write out selected parcel features to file (SHP)

    # 6. Write out selected street attributes to a table file (DBF)

except:
```

10. Add the Select Layer by Location line.

Add the Select Layer by Location line (4b). The default spatial relationship "INTERSECT" is used as well as the default selection method "NEW_SELECTION" since this is a new selection on the parcel layer. Make sure to use the parcel layer (poly_layer) as the Input Feature Selection and the street layer (feat_layer) as the select features layer. The user can "hard code" the search distance parameter ('200 FEET') or can create a variable that is assigned to the character string ('200 FEET'). The author has chosen the latter which allows a convenient way to change the search distance (and it is near the top of the script with other variable definitions).

```
search_distance  =  '200 FEET'
```

Notice that the line continuation character ("\") is used, since the Select Layer by Location line is a little longer that the other lines. This character makes it convenient to keep long lines compact to make the code easier to read. Also note that both single and double quotes are used. Since the parameters are strings, this is ok.

```
import arcpy, sys, traceback

arcpy.env.workspace = "C:\\PythonPrimer\\Chapter05\\Data\\"

feat_class = "Sacramento_Streets.shp"
poly_fc = "City_Parcels.shp"

feat_layer = "Streets"
poly_layer = "City Parcels"

search_distance = '200 FEET'

try:

    # 1. Make a feature layer for streets
    arcpy.MakeFeatureLayer_management(feat_class, feat_layer)

    # 2. Create a query statement to select highways

    # Create query statement
    query = '"CLASS" = \'H\''

    # 3. Select by attributes using the query

    # Select features by attribute using query
    arcpy.SelectLayerByAttribute_management(feat_layer, "NEW_SELECTION", query)

    # 4. Select parcels that are 200 ft from highways

    # 4a. Make feature layer
    arcpy.MakeFeatureLayer_management(poly_fc, poly_layer)

    # 4b. Select Parcels
    arcpy.SelectLayerByLocation_management(poly_layer, "INTERSECT", \
                                feat_layer, search_distance, "NEW_SELECTION")
```

11. Count the number of records in each selection and print to the Python Shell (4c).

It might be a good idea to make sure that our selection code is selecting the expected number of records. The `GetCount` method can be used to obtain the number of selected records and assign them to a variable. Subsequently, this variable can be used in a print statement to report back to the code developer the number of records to the Python Shell.

Add the `GetCount` code for both feature layers and a print statement. Note that the "result" variable is "cast" (i.e. converted) to a string so that it can be properly concatenated with the other text in the print statement.

Note the use of the new line character ("\") and the new line feed continuation character ("\n") in the print statements. The new line feed continuation character will break the long print statement into two lines. The "pseudo-code" has also been updated.

```
# 1. Make a feature layer for streets

arcpy.MakeFeatureLayer_management(feat_class, feat_layer)

# 2. Create a query statement to select highways

# Create query statement
query = '"CLASS" = \'H\''

# 3. Select by attributes using the query

# Select features by attribute using query
arcpy.SelectLayerByAttribute_management(feat_layer, "NEW_SELECTION", query)

# 4. Select parcels that are 200 ft from highways

# 4a. Make feature layer

arcpy.MakeFeatureLayer_management(poly_fc, poly_layer)

# 4b. Select Parcels
arcpy.SelectLayerByLocation_management(poly_layer, "INTERSECT", \
                                       feat_layer, search_distance, "NEW_SELECTION")

# 4c. Get the number of selected records and create a print statement for each

# print the number of selected records in the feature layer with a query
result = arcpy.GetCount_management(feat_layer)
print "Number of selected features in the feature layer " + feat_layer + \
      " with a query: "+ str(result)

result = arcpy.GetCount_management(poly_layer)
print "Number of selected features in the feature layer " + poly_layer + \
      "\n within a distance of " + search_distance + ": "+ str(result)
# 5. Write out selected parcel features to file (SHP)

# 6. Write out selected street attributes to a table file (DBF)
```

12. Write out the selected parcels to a new feature class (**out_parcels.shp**).

Add the Copy Features line to the code to write out the selected parcels to a new feature class. A variable has been defined at the top of the code (above the `try:` block) to the name of the output shapefile (**out_parcels.shp**). A print statement has also been added to report that the file has been created.

```
out_poly_fc = "out_parcels.shp"
```

```
# 4b. Select Parcels
arcpy.SelectLayerByLocation_management(poly_layer, "INTERSECT", \
                                    feat_layer, search_distance, "NEW_SELECTION")


# 4c. Get the number of selected records and create a print statement for each

# print the number of selected records in the feature layer with a query
result = arcpy.GetCount_management(feat_layer)
print "Number of selected features in the feature layer " + feat_layer + \
      " with a query: "+ str(result)

result = arcpy.GetCount_management(poly_layer)
print "Number of selected features in the feature layer " + poly_layer + \
      "\n within a distance of " + search_distance + ": "+ str(result)
# 5. Write out selected parcel features to file (SHP)

# Copy the selected polygon features to a new shapefile

arcpy.CopyFeatures_management(poly_layer, out_poly_fc)
print "Copied selected features from " + poly_fc + " to " + out_poly_fc

# 6. Write out selected street attributes to a table file (DBF)
except:
```

13. Write out the selected street attributes to a new standalone table (**out_street_attributes.dbf**).

In this case, only the selected attributes (not the geographic features) are written to a standalone table (a dBase formatted table). A variable is added to the top of the script to hold the value of the output table name. A `print` statement is added to report that the output table has been created.

```
out_street_attributes = "out_street_attrib.dbf"
```

```
result = arcpy.GetCount_management(feat_layer)
print "Number of selected features in the feature layer " + feat_layer + \
    " with a query: "+ str(result)

result = arcpy.GetCount_management(poly_layer)
print "Number of selected features in the feature layer " + poly_layer + \
    "\n within a distance of " + search_distance + ": "+ str(result)
# 5. Write out selected parcel features to file (SHP)

# Copy the selected polygon features to a new shapefile

arcpy.CopyFeatures_management(poly_layer, out_poly_fc)
print "Copied selected features from " + poly_fc + " to " + out_poly_fc

# 6. Write out selected street attributes to a table file (DBF)

# Copy the selected street segment attributes to a dBase table

arcpy.CopyRows_management(feat_layer, out_street_attributes)
print "Copied selected attributes from " + feat_class + " to " + out_street_attributes

print "Completed Script"

except:
```

Notice that a `print` statement is written indicating that the script is complete.

At this point, the script is complete. The code developer can Save and Check the script. Fix any syntax errors that occur. Make sure to review the ArcGIS Help for any of the functions and use ArcMap to assist in building the proper query syntax.

If the user attempts to run the script more than one time, it will produce an error because the output files already exist. Some if statements using the `Exists` and `Delete` functions can help alleviate this problem.

`If` statements using the `Exists` and `Delete` functions can be used for feature classes, feature layers, tables, and table views, among others). The screen shot below shows several places in the demo script that uses these methods. Refer to the **Demo 5b.py** script for all placements of the `Exists/Delete` routines.

```
# 4a. Make feature layer

if arcpy.Exists(poly_layer):
    arcpy.Delete_management(poly_layer)

arcpy.MakeFeatureLayer_management(poly_fc, poly_layer)


# 4b. Select Parcels
arcpy.SelectLayerByLocation_management(poly_layer, "INTERSECT", \
                                       feat_layer, search_distance, "NEW_SEL


# 4c. Get the number of selected records and create a print statement for ea

# print the number of selected records in the feature layer with a query
result = arcpy.GetCount_management(feat_layer)
print "Number of selected features in the feature layer " + feat_layer + " w

result = arcpy.GetCount_management(poly_layer)
print "Number of selected features in the feature layer " + poly_layer + \
      "\n within a distance of " + search_distance + ": "+ str(result)

# 5. Write out selected parcel features to file (SHP)

# Copy the selected polygon features to a new shapefile

if arcpy.Exists(out_poly_fc):
    arcpy.Delete_management(out_poly_fc)

arcpy.CopyFeatures_management(poly_layer, out_poly_fc)
print "Copied selected features from " + poly_fc + " to " + out_poly_fc

# 6. Write out selected street attributes to a table file (DBF)


# Copy the selected street segment attributes to a dBase table

if arcpy.Exists(out_street_attributes):
    arcpy.Delete_management(out_street_attributes)
```

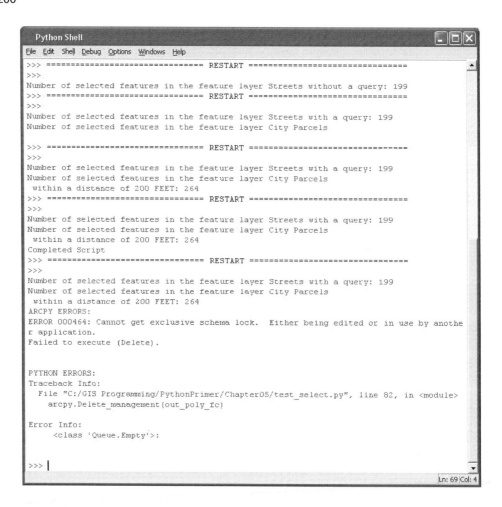

The figure above shows an example of an error code if the existing data cannot be deleted.

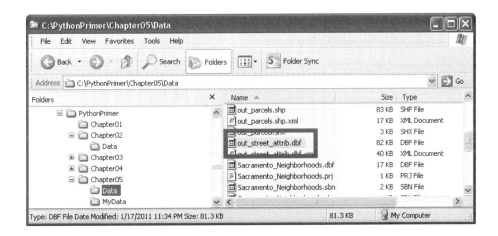

The figure above shows the folder with the output dBase file (**out_street_attrib.dbf**). The reader's folder for the output may be different.

The **Demo5 script results.doc** document can be compared with the reader's results for this demo.

Exercise 5: Select Features by Attribute/Location and Write them to a New Feature Class

Exercise 5 will provide the opportunity for the code developer to put into practice some of the concepts illustrated in Chapter 5.

Write a Python script with the following conditions. Use the demo scripts, chapter content, ArcGIS Help, and Python resources as needed.

Review the **Sacramento_Streets.shp** file and **Sacramento_Neighborhood.shp** file in the **Exercise5.mxd**. You will only need to use the ArcMap Document and look at the data and attributes for your review. You will not need the ArcMap document to write the script.

Write a script with the following conditions. The goal is to have a new feature class that only contains the local streets that are located within a neighborhood of your choice.

1. Create a selection on the Neighborhoods shapefile that uses a query variable that queries one of the neighborhoods using the NAME field that is covered by the Sacramento Streets file. Simply use one of the Neighborhood names in the query. You should not have to programmatically select a neighborhood that is covered by the streets.

2. Using the Selected neighborhood from above, use it to "spatially select" the streets within the neighborhood.

3. Using a query, select only the "Local" streets from within this neighborhood. The query will need to use the "CLASS" attribute in the streets data.

4. Write a print statement that indicates the number of selected streets

5. Write these selected street features to a new feature class in the **\MyData** folder. Use the shapefile format.

6. Make sure to use variables, queries, and feature layers appropriately

7. Make sure to use Exists and Delete functions as necessary

8. Use the `try` block and exception code from previous exercises

Extra

Do the same as above, but for the **Z'berg Park** neighborhood and collector street types (CLASS value is 'C' for collectors). Write out a feature class and a separate standalone table of the selected street features.

The author strongly recommends that the Exercise is working first before attempting the Extra section.

Chapter 5: Questions

1. What is the difference between the following:

    a. Feature Class
    b. Feature Layer
    c. Table
    d. Table View

2. What are the following used for with developing queries?

    a. % (non numeric)
    b. * (non numeric)
    c. LIKE
    d. AND
    e. OR

3. How are NULL values used in query strings? Give a query example using querying for NULL values.

4. What Toolbox and Toolset can the Select Layer tools and Make Feature Layer or Make Table View Tools be found?

5. What is the default selection type for the Select Layer By Attribute?

6. What is the default selection type for the Select Layer By Location?

7. Can a feature class be used as the input in a Select Layer By Attribute or Select Layer By Location? If no, what must be used?

8. What is the geoprocessing routine (tool) to create a new feature class from the selected features in a feature class?

9. What is the geoprocessing tool to create a new table from selected records in a table?

Chapter 6 Creating and Using Cursors and Table Joins

Overview

Cursors are common database constructs that are used to access, read, and update values in a database table (or in the case of GIS, a feature class attribute table). One can think of cursors as a means to "point to" a collection of data records (features or rows in a table) from which the programmer can systematically process individual records. For example, suppose a GIS analyst performed an attribute selection on a number of records or features and then wanted to apply a *different* value to one of the attributes (columns) or read a value from one of the attributes to then use as an input for a query or copy it into a different table. Performing this action by using the Select Layer by Attribute and Calculate Field functions can be time consuming and repetitive within ArcMap. Even in a Python script the selection and calculate tasks would need to be re-used for each unique combination of selection criterion and field calculation. But, by using cursors, a more logical series of steps can be performed allowing reading from and/or updating multiple attribute fields for a given record (or feature).

"Table joins," attribute tables from two different sources that can be connected to one another through a common attribute, is also a concept that can be discussed when using cursors, since cursors are often used to read and write values from a variety of data sources as well as to create and summarize data from these sources. The latter part of this chapter will focus on how to programmatically create and use "joined" tables.

Cursors

The general concept of a cursor is to access one or more records from a database (tabular or geographic). Once this set of records is obtained, each record can be acted upon. The three actions that can be performed in ArcGIS are:

1. *Search* – read values from a record
2. *Insert* – create new records
3. *Update* – modify existing records

The search and update cursors also have a query parameter ("where clause") that can be used to access a subset of data. If the query parameter is not used, then the cursor will operate on all of the records in a dataset.

The cursor is different from a "selected" set of records in such that the records in a cursor are not necessarily "selected" where the selected records would appear highlighted in ArcMap. In addition, the records in a cursor are not going to be used in a spatial overlay such as a SelectByLocation operation.

Given a set of records (features or tabular records) the cursor does the following:

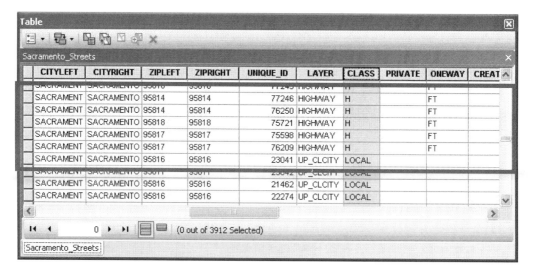

In the figure above the box indicates the group of records accessed by a search cursor that uses a query parameter where "CLASS" = 'H'. The figure below shows the Python syntax for a search cursor that uses a query parameter. Refer to the general_cursor_sample.py script that can be found under the **\PythonPrimer\Extra_Samples** folder.

```
general_cursor_sample.py - C:/PythonPrimer/Extra_Samples/general_cursor_samp...
File  Edit  Format  Run  Options  Windows  Help

# assign a variable to hold the street shapefile
streets_shp = 'Sacramento_Streets.shp'

try:

    # set a variable to hold the query string
    query = '"CLASS" = \'H\''

    # Create a search cursor to access rows that
    # have a road class of "H" - Highway

    # Gets a collection of rows from a feature class or table
    srows = arcpy.SearchCursor(streets_shp, query)
```

Once the collection of records is retrieved, individual rows (features or records) can be processed. The figure below shows the syntax to iterate through the records accessed by the cursor.

```
general_cursor_sample.py - C:/PythonPrimer/Extra_Samples/general_cursor_sam...
File  Edit  Format  Run  Options  Windows  Help

# assign a variable to hold the street shapefile
streets_shp = 'Sacramento_Streets.shp'

try:

    # set a variable to hold the query string
    query = '"CLASS" = \'H\''

    # Create a search cursor to access rows that
    # have a road class of "H" - Highway

    # Gets a collection of rows from a feature class or table
    srows = arcpy.SearchCursor(streets_shp, query)

    for srow in srows:    # for each row in the cursor

        # assign a variable for the value of srow.FULLSTREET
        # assign a variable for the value of srow.getValue("UNIQUE_ID")
        fullstreet = srow.FULLSTREET
```

A `for` loop is used to iterate through the collection of records. In the example above the script finds and obtains the "FULLSTREET" attribute value for the specific record and then prints it to the screen. Note that the `for` loop uses a variable `srow` to access a specific row from the collection of rows (`srows`), the cursor. `srows` is a variable that accepts the results of the search cursor.

As described in Chapter 2, `for` loops are used to iterate over elements in a collection of objects. The `for` loop is constructed by using the general syntax below.

```
for <specific_elements> in <a_collection_of_objects>:
```

In ArcGIS the for loop is often used to iterate over the table rows, features of a geographic file, and specific elements of the ArcGIS "List" routines (See Chapter 7 and the ArcGIS Help on working with ArcGIS lists).

NOTE: In a more general sense for loops can also iterate over numbers in a sequence or specific items in a Python list. See Chapter 3, a Python text, or the python.org site for more details on Python lists.

In order to access a specific element in a collection, a variable is used (e.g. `srow`) within the `for` loop. See the illustration above.

The record outlined by the box below represents the first record (i.e. the specific row), when the `for` loop is initially processed by Python. Python interprets the `for` loop line as "get a row" (`srow`) from the collection of rows (`srows`).

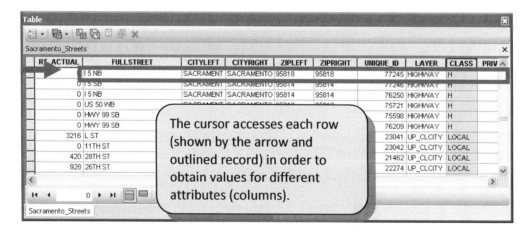

The reader should note that when implementing cursors in Python, there is no "selection" done, but rather a method to access records for further processing.

One can think of the cursor as an in-memory function to access data and then process it.

The above example shows using a feature class directly in a cursor. Feature layers, tables, and table views can also be used with cursors. See Chapter 5 for creating and using feature layers and table views.

Types of Cursors

As already mentioned, three kinds of cursors exist in ArcGIS and are themselves ArcGIS routines. Depending on the type of cursor, the code developer can read values from features or tables, create new records and set initial values, or update existing records with new attribute values.

1. SearchCursor – for reading values from a set of records
2. InsertCursor – for creating and setting initial values of a record
3. UpdateCursor – for updating existing records in a table or feature class

This chapter will focus on cursor methods created by Esri and used in ArcGIS which can be programmed using Python. The reader can refer to other databases and GIS functionality in Appendix 5.1.

Implementing Cursors

Cursors in ArcGIS begin with implementing one of the three kinds of cursors mentioned above. Each of the three methods will be illustrated below. The reader will then have the opportunity to write a Python script that uses all three. The reader can refer to the following ArcGIS Help for more details **Geoprocessing - Geoprocessing with Python - Accessing geographic data in Python.** Each of the cursor types will be illustrated below.

Search Cursor

The `SearchCursor` is used to "read" and obtain data values from a table or feature class. Typical uses of the search cursor can include "pulling" (reading from) data values from one table or feature class to then put them in another table or feature class. One might think that using the `SelectLayerByAttribute` using a query and setting some visibility parameters for hiding or showing certain fields from an attribute table is possible. This is possible, but this method may likely require a number of different steps and queries to be performed and/or intermediate feature classes to be created. An alternative method can use the `SearchCursor` to access specific rows and then specific attributes (columns) to "get" the data and then use another cursor type to write or "set" the rows and values in another table. A common implementation of such method can be used when data needs to be extracted from an enterprise grade database table (such as an Oracle ArcSDE feature class) and the resulting records and values are written to a file or personal geodatabase or dBase file for another user that does not have access to an ArcSDE environment or enterprise database or the resulting table may need to be sent to a 3rd party organization that does not have direct access to an organization's database infrastructure.

Creating and Using the Search Cursor

The first step for using a `SearchCursor` is to access a feature class or table. The code developer needs to supply a workspace and possibly a variable to hold the feature class or table name. Since cursors can work with feature layers or table views, the Make Feature Layer or Make Table View can be used, if required. The search cursor contains the following parameters. Only the first parameter is required. For a full set of parameters, see the ArcGIS Help for Search Cursor. Optional parameters are shown with {}.

```
SearchCursor(<dataset>, {where clause}, {spatial
reference}, {fields}, {field sort})
```

dataset – name of the feature class, table, feature layer, or table view *where clause* – query statement to limit the number of records in the collection of records

spatial reference – the name of the coordinate system (if it is needed). If the feature class has a coordinate system, this parameter may not be needed.

fields – a list of fields separated by a semi-colon (;). The list can be a partial list of available fields

field sort – uses the structure <FIELD_NAME SORT_TYPE>. The field name and sort type are separated by a space. The sort type is either (A) for ascending or (D) for descending. Multiple fields can be listed to sort one column and then another column. The fields will be separated by a (;). For example, multiple fields may be sorted in a cursor like the following:

"STREET_NAME A; STREET_NUM A"

The street name attribute is sorted and then the street number. A resulting table may have a list of street address sorted by the street name (A to Z) and then the number with the order for each street going from a low to high address number.

The following figure illustrates the creation of the search cursor using a query to limit the returned records that include only highway street segments.

```
import arcpy, sys, traceback

# set the current workspace (in the case a folder)
arcpy.env.workspace = 'C:\\PythonPrimer\\Chapter06\\Data\\'
# assign a variable to hold the street shapefile
streets_shp = 'Sacramento_Streets.shp'

try:

    # set a variable to hold the query string
    query = '"CLASS" = \'H\''

    # Create a search cursor to access rows that
    # have a road class of "H" - Highway

    # Gets a collection of rows from a feature class or table
    srows = arcpy.SearchCursor(streets_shp, query)
```

Note that the variable `srows` is used to hold the results (a collection or records) of the `SearchCursor` (right side of the equal sign).

After the records are retrieved from the dataset and assigned to the cursor, the individual records (rows) are accessed to retrieve specific values from the attributes (columns). A for loop is used to iterate and access each row.

Using the *for* Loop

A `for` loop that operates over a typical search cursor looks like the following figure.

```
try:
    # set a variable to hold the query string
    query = '"CLASS" = \'H\''

    # Create a search cursor to access rows that
    # have a road class of "H" - Highway

    # Gets a collection of rows from a feature class or table
    srows = arcpy.SearchCursor(streets_shp, query)

    for srow in srows:    # for each row in the cursor

        # assign a variable for the value of srow.FULLSTREET
        # assign a variable for the value of srow.getValue("UNIQUE_ID")
        fullstreet = srow.FULLSTREET
        unique_id = srow.getValue("UNIQUE_ID")

        # prints the value of the variables
        # i.e. the value of srow.FULLSTREET and srow.GetValue("UNIQUE_ID")
        print fullstreet + ' ' + str(unique_id)
```

Notice the general structure of the `for` loop.

```
for <item> in <items>:
    # indent code to process as a block
```

As described earlier, the `for` loop sets up the structure to retrieve an element (for or feature in this case) from a collection of rows or features. A variable (`srow`, in this example) will be used to reference a specific row <item> and <items> will be replaced by the variable used to reference the cursor (`srows`, in this example).

Once the `for` loop line is interpreted by Python, individual attribute values can be retrieved from the respective attribute column. In addition, other processes can be added inside the `for` loop. Notice that the lines of code that make up the `for` loop "block of code" are indented which is required for a looping structure in Python. Indentation tells Python to process the subsequent lines of code as a block. When the last line of the block within the `for` loop is completed, other code can be "un-indented" to perform other operations.

Generally, a specific value is obtained from a column at the given row to use with other operations in the script. The general method to achieve this is to write the syntax:

```
row.FieldName
```

where `row` is the variable used in the for loop to access a specific row from the cursor and `FieldName` is the specific name of an attribute field (column). For a given row and column in a table, the value for a (cell) can be retrieved.

In the above example, the column FULLSTREET is accessed for the given row:

```
srow.FULLSTREET
```

`srow` is a variable that holds a pointer to a specific row in the collection of rows (`srows`). The right side of (`srow.`) indicates the column (attribute or field name). For example, if the name of a street is "Main St" that is stored in the column FULLSTREET for the given row, then `srow.FULLSTREET` will hold the value "Main St". The field name is case sensitive. Also, if the field contains spaces, this syntax will not work.

NOTE: It is recommended that field names do not include spaces, dashes, special characters, or begin with a number; underscores are acceptable.

An alternative method for accessing a column is to use this syntax. The syntax of `getValue` must be used. See Cursors under ArcGIS Help and look for getValue.

> `row.getValue("FieldName")`

In this example, the syntax is:

> `srow.getValue("UNIQUE_ID")`

This method can accommodate field names with spaces, but again, it is highly recommended that field names do not have spaces, dashes, or special characters. Some database formats may not allow such characters and may influence how databases and workflows are designed.

The results of the search cursor example above, prints the full street name and unique ID number to the Python Shell. In a practical sense, these values will be used in other parts of the script.

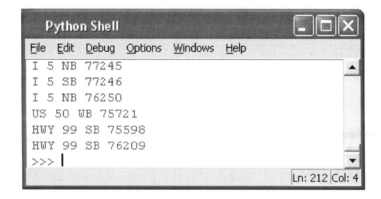

Using the *while* Loop

A second method to process rows of a cursor is to use a `while` loop. The code for the `while` loop functions differently than the `for` loop.

```
while <condition is true>:
    # indent code to process as a block
```

To perform the same operation as described above using the `while` loop, the code looks like the following:

```python
# set the current workspace (in the case a folder)
arcpy.env.workspace = 'C:\\PythonPrimer\\Chapter06\\Data\\'
# assign a variable to hold the street shapefile
streets_shp = 'Sacramento_Streets.shp'

try:

    # set a variable to hold the query string
    query = '"CLASS" = \'H\''

    # Create a search cursor to access rows that
    # have a road class of "H" - Highway

    # Gets a collection of rows from a feature class or table
    srows = arcpy.SearchCursor(streets_shp, query)


    srow = srows.next()

    while srow:    # while a row exists in the cursor

        # assign a variable for the value of srow.FULLSTREET
        # assign a variable for the value of srow.getValue("UNIQUE_ID")
        fullstreet = srow.FULLSTREET
        unique_id = srow.getValue("UNIQUE_ID")

        # prints the value of the variables
        # i.e. the value of srow.FULLSTREET and srow.GetValue("UNIQUE_ID")
        print fullstreet + '  ' + str(unique_id)

        srow = srows.next()
```

Three changes are required to use the `while` loop to process a cursor.

1. A line that implements the `.next()` method is needed to gain access to the first "actual row" of the cursor.
2. The `while` loop line is required that checks to see if a rows exists in the cursor (collection of rows).
3. At the end of the `while` block, the same `.next()` method is required to access the next row in the cursor (if one exists).

Unlike the `for` loop, a `while` loop requires that a row be accessed. The `srow = srows.next()` line performs this operation. One can think of the cursor as a closed box that contains objects in it. Before the objects (rows) can be taken out of the box (the cursor), the box must be opened and then the object on top (the first row) can be chosen. The `.next()` method before the while loop accomplishes this. The `while` loop performs this check. If the row exists, then the `while` block code is processed; if a row does not exist, then the `while` loop is by-passed and other code is processed. The `.next()` method above the `while` loop accesses the first row from the cursor.

To process other rows (records) in the cursor, the `.next()` statement at the end of the `while` block is required. (Note that this `.next()` statement is indented relative to the `while` statement). This line moves the pointer from one row to the next row in the cursor (i.e. the next object from the box is chosen). The `while` statement then checks to see if this row exists in the cursor. If it does, then the `while` block code is processed. If a row does not exist, then the `while` block code is by-passed and other code is processed.

The code above using the `while` block produces the same result as the `for` loop. Either method is valid, however, the `for` loop requires two fewer lines. The code developer should follow proper Python syntax and coding structures (e.g. indentation or the `.next()` method) correctly when using either method for looping over rows in a cursor.

Insert Cursor

The second kind of cursor is the `InsertCursor`. An insert cursor allows a code developer to create new rows for a table or feature class and assign initial values to one or more of the attributes. A prerequisite for using the `InsertCursor` is a table or feature class must exist that contains attributes (columns). The code below illustrates using the `CreateTable` ArcGIS function to create a file geodatabase table from scratch and use the `InsertCursor` to add rows and populate some of the initial values for the attributes. See the ArcGIS Help for `CreateTable` and `AddField` routines.

```
# set the current workspace (in the case a file geodatabase)
arcpy.env.workspace = 'C:\\PythonPrimer\\Chapter06\\Data\\cursors.gdb'

outpath = 'C:\\PythonPrimer\\Chapter06\\Data\\cursors.gdb'

# assign a variable to hold the address table
address_table = 'addresses'


try:

    # check to see if the address table exists
    # if it does, delete it
    if arcpy.Exists(address_table):
        arcpy.Delete_management(address_table)

    # create a new table in the given location, in this case
    # a file geodatabase
    arcpy.CreateTable_management(outpath, address_table)

    # add fields to the table
    arcpy.AddField_management(address_table, 'AddID', 'LONG')
    arcpy.AddField_management(address_table, 'StreetNum', 'LONG')
    arcpy.AddField_management(address_table, 'StreetName', 'TEXT', '', '60')

    print 'Created table and added fields'
```

The code above shows how to create a file geodatabase table from scratch. An if statement is used to check to see if a table already exists in the geodatabase. If it does, then it is deleted, since the next step actually creates the table. The `CreateTable` routine is used to create an empty table (**addresses**) and then the `AddField` routine is used to create new attributes. Notice the format of the `AddFields` routine and the use of data types. See the ArcGIS Help for additional information on defining and formatting attribute fields.

Running the script up to this point creates a new table in the file geodatabase called **addresses** with the added fields. If ArcCatalog is opened, the following should appear when the user navigates to the location of the new table. If the reader is following along writing their own code, see the Schema Lock section and note below.

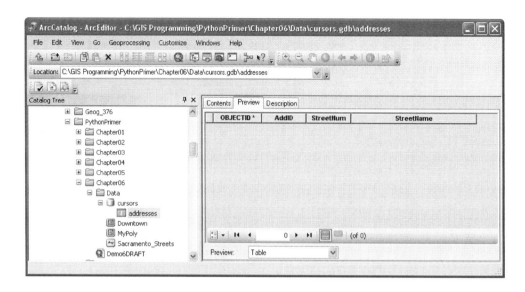

In the figure above under the Preview tab note that the attribute fields are present. In addition, the OBJECTID field is added. This is automatically done by ArcGIS and must be present for GIS operations within ArcGIS. Tables or feature classes that do not have this field may not be able to perform overlay operations, table joins, or create data from attributes that have Latitude/Longitude (or X/Y fields). No records (or rows) have been added. The Insert Cursor will be used to do this.

Schema Locks on Data

The reader may have already experienced challenges with deleting data when actively writing a Python script. The code developer may have seen a similar message shown below indicating a schema lock.

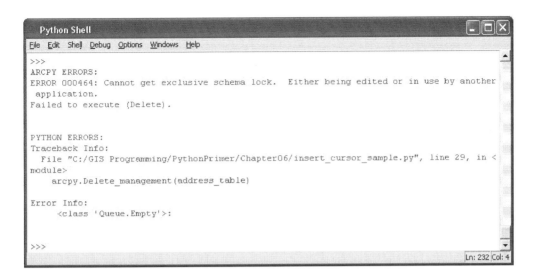

In the case above a Python script is actively being developed and run from Python IDLE. To help troubleshoot syntax and processing problems, the code developer decided to open ArcCatalog to check the results of an operation. Subsequently, the code developer made additional changes in the open Python script, saved the edits, run the Check Module, and attempted to re-run the script. Because both ArcCatalog and Python were open at the same time, a lock (schema lock) was put on the data. When the Python script attempted to delete the table, the error message was produced.

The remedy for such problem is to make sure all ArcGIS applications are closed completely. If Python is still open, the user can attempt to run the code again.

NOTE: In some cases, Python IDLE and the Python Shell must be completely closed in addition to any ArcGIS application because the Python program can access data. Oftentimes, this kind of problem occurs with cursors, looping functions, and attempting to update values that may or may not be available. The general rule to remedy this problem is to close all ArcGIS applications, Python Shell, and Python IDLE. Make sure to save changes to the Python script before closing. Most of the time a user does not need to shut down or restart a machine. Performing the above tasks should eliminate most schema locks.

Creating and Using the Insert Cursor

Now that a table is available, some records can be added to the table as well as initialize some of the values.

Before adding records (rows), the insert cursor must be defined. A variable `irows` is used to hold the results of the insert cursor.

```
irows = arcpy.InsertCursor(<dataset>, {spatial reference})
```

dataset – is the name of the table or feature class
spatial reference – is an optional parameter to define the coordinate system, if required.

```
# assign a variable to hold the address table
address_table = 'addresses'

try:

    # check to see if the address table exists
    # if it does, delete it
    if arcpy.Exists(address_table):
        arcpy.Delete_management(address_table)

    # create a new table in the given location, in this case
    # a file geodatabase
    arcpy.CreateTable_management(outpath, address_table)

    # add fields to the table
    arcpy.AddField_management(address_table, 'AddID', 'LONG')
    arcpy.AddField_management(address_table, 'StreetNum', 'LONG')
    arcpy.AddField_management(address_table, 'StreetName', 'TEXT', '', '60')

    print 'Created table and added fields'

    # Create cursor object
    irows = arcpy.InsertCursor(address_table)
```

Once the insert cursor is defined, a looping structure can be used to add and set rows in the new table.

```
arcpy.AddField_management(address_table, 'StreetNum', 'LONG')
arcpy.AddField_management(address_table, 'StreetName', 'TEXT', '', '60')

print 'Created table and added fields'

# Create cursor object
irows = arcpy.InsertCursor(address_table)

# counter variable to initialize loop
counter = 1

# run the loop until counter is <= to 10
while counter <= 10:

    irow = irows.newRow()      # create a new row
    irow.AddID = counter       # update the AddID field
                               # by calculating it to
                               # the counter value

    irows.insertRow(irow)      # actually inserts the row into the table
    counter += 1               # increment the counter so a a new row can be added

del irow, irows    # delete the cursor varaible to prevent schema locks

print 'Added ' + str(counter - 1) + ' rows to table: ' + address_table

except:
```

The `while` loop is used to check the conditions so that the while block code can be processed, in this case, adding and setting rows in a table. The first line in the `while` block creates the new row (`irows.newRow`). The next line illustrates one method to "set" an initial value for an attribute (AddID), in this case the value of the counter variable. If an attribute was a string, then a specific name can be used. For example, if records for highways are added, a value such as `row.RoadType = 'Highway'` might be used to set the RoadType attribute to be the string 'Highway'. In addition, a calculation may be needed on the right side of the equal sign.

NOTE: This example shows a `while` loop being used to create new rows. A loop is not always required to add (insert) new rows to a table. For example, a search cursor can be set up to read rows from one table and the insert cursor can be created where a new row is inserted into a separate standalone table. In this case the new row and the insertion of the row are performed within the loop for the search cursor. From a code development point of view, it is up to the developer to determine what kind of looping structure is required and when to implement different functions. This is part of the challenge the code developer faces when writing scripts. Detailed workflows become very important in these instances.

An alternative method to "set" the value of an attribute can use the following syntax. See the ArcGIS Help for getting and setting values. Either method or combination is valid. It is recommended that one method is used consistently to develop code.

```
irow.setValue("AddID", counter)
```

Using the `setValue` method does not involve the use of an equal (=) sign, but contains the name of the output field "AddID" and the value (e.g. the value assigned to the `counter` variable in this example) that "AddID" will be set to.

Notice that the other attributes are not set because the initial values are not known. Other methods will likely be used to update them.

The next line,

```
irows.insertRow(irow)
```

actually inserts the row (`irow`) into the table via the cursor (`irows`). The last line of the block increments the `counter` variable by 1 so the `while` loop can check the condition to see if it is true. A new row is inserted and updated until the `while` loop condition is false (i.e. the `counter` variable is greater than 10).

Outside of the loop a Python `del` statement is used to "free up" the memory the irow and irows variables are using and helps to eliminate one source of data locking.

Once the insert cursor has created new rows, the script up to this point should produce the following results. ArcCatalog will be needed to see the table.

If the reader is writing the code while reading this section, the code can be saved and the script run. Navigate to the location of the table and open it in ArcCatalog. Make sure to close Python IDLE and the Python Shell before doing so.

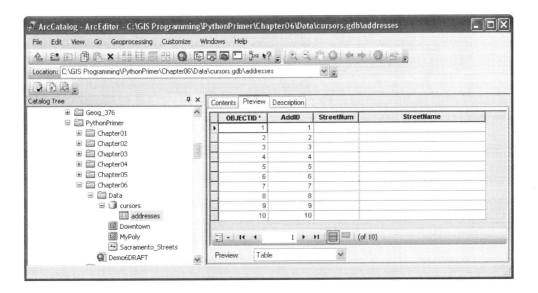

Update Cursor

The `UpdateCursor` can be used to update existing values in a table or feature class. Similar to the search cursor a query, spatial reference, field list, and field sorting can be used as parameters.

```
urows = arcpy.UpdateCursor(<data set>, {where clause},
{spatial reference}, {field list}, {sort fields})
```

To illustrate how values from one table (attribute table) can update values in a separate table, both the search cursor and update cursor will be used. This example is commonly used to read data from one data set and put it into another data set for use in other systems or organizations. This type of activity might be automated to run weekly to produce an updated table that another analyst, user, or external source can use.

The script begins with both a workspace and output data paths (`fgdpath`) defined.

```python
import arcpy, sys, traceback

# set the current workspace (in the case a folder)
arcpy.env.workspace = 'C:\\PythonPrimer\\Chapter06\\Data\\'

fgdpath = 'C:\\PythonPrimer\\Chapter06\\Data\\cursors.gdb\\'

address_shp = 'addresses.shp'
address_fgd = fgdpath + 'addresses'

try:

    # set a variable to hold the query string
    query = '"FID" >= 0 AND "FID" <= 10'


    # Gets a collection of rows from a feature class or table
    srows = arcpy.SearchCursor(address_shp, query)
```

Notice the path for the workspace. The workspace is a folder that contains the shapefile (**addresses.shp**) from which the search cursor will read data values from the address shapefile's attribute table. A separate path variable (`fgdbpath`) is defined to the file geodatabase location where the updated feature class will reside when the script is executed. This geodatabase contains a pre-existing table (**addresses**) to which rows will be added using the update cursor.

After the data path definitions, a query is created to limit the number of records pulled (i.e. accessed) from the address shapefile. The query is returning only the first ten records from the address file using the FID attribute to match the number of records in the output table and to illustrate the use of the search and update cursors. In practice with a case like this, a different kind of query such as only return "residential" addresses vs. "business" addresses might be used.

The next step is to start the looping structure to cycle through and read each row in the search cursor.

```
try:

    # set a variable to hold the query string
    query = '"FID" >= 0 AND "FID" <= 10'

    # Gets a collection of rows from a feature class or table
    srows = arcpy.SearchCursor(address_shp, query)

    obj_id = 1

    for srow in srows:

        # assign a for the value of srow.STREETNAME
        # assign a variable for the value of srow.getValue("STREETNUMB")
        streetname = srow.STREETNAME
        streetnum = srow.getValue("STREETNUMB")
```

The STREETNAME and STREETNUMB attribute values are read and stored in a variable. Note the `obj_id` variable set to 1. This will be used in the next step.

Creating and Using the Update Cursor

Since both the `SearchCursor` and `UpdateCursor` function row by row (or record by record), the code developer needs to devise a strategy to get the correct data from one table and put it in the correct location in the update table. In a sense, the rows in each table are synchronized. Working out the logic for this kind of task is challenging and may require the code developer to write out in detail the specific steps that need to occur for each dataset. Also, the code developer can expect some "trial and error" to occur to make sure the correct information is being read from and written to the right location in each data set. Some clues that indicate the logic is not right can include:

1. The same value being written out to each row
2. Only the first or last value being written out to each row
3. Empty values appearing where data is expected to be written to the output row values

4. Out of range errors, which indicate that the looping mechanisms are not iterating or incrementing properly
5. Queries that use incremented values (such as the use of FID and `obj_id` above) return values that do not exist or go beyond the physical range of the table (e.g. a table may only have 100 records, but the query indicates there are 101 records).

If errors occur, it is helpful to systematically work through each piece of code rather than try to work out the entire problem at once.

The `UpdateCursor`, in this case, will use a query to access a single record in the table that will be updated.

```
query = '"OBJECTID" = ' + str(obj_id)
```

```
# Gets a collection of rows from a feature class or table
srows = arcpy.SearchCursor(address_shp, query)

obj_id = 1

for srow in srows:

    # assign a for the value of srow.STREETNAME
    # assign a variable for the value of srow.getValue("STREETNUMB")
    streetname = srow.STREETNAME
    streetnum = srow.getValue("STREETNUMB")

    # Get a collection of rows for the update cursor
    # based on a query
    # This update cursor will only return a single
    # record, which is desired.
    query = '"OBJECTID" = ' + str(obj_id)
```

The OBJECTID attribute can be found in the file geodatabase table (addresses). Notice that the first record has an OBJECTID value of 1 and subsequent records are ordered sequentially.

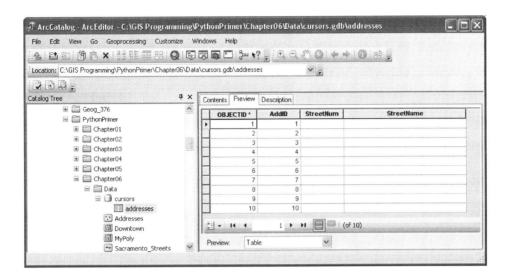

The query is used as a parameter in the update cursor statement. Notice that the variable (`address_fgd`) pointing to the file geodatabase table (**addresses**) is the dataset for the update cursor.

```python
    # Gets a collection of rows from a feature class or table
    srows = arcpy.SearchCursor(address_shp, query)

    obj_id = 1

    for srow in srows:

        # assign a for the value of srow.STREETNAME
        # assign a variable for the value of srow.getValue("STREETNUMB")
        streetname = srow.STREETNAME
        streetnum = srow.getValue("STREETNUMB")

        # Get a collection of rows for the update cursor
        # based on a query
        # This update cursor will only return a single
        # record, which is desired.
        query = '"OBJECTID" = ' + str(obj_id)

        # create the update cursor
        urows = arcpy.UpdateCursor(address_fgd, query)

        # cycle through the rows (in this case only 1)
        # to actually update the row in the cursor
        # with the values obtained from the search cursor
        for urow in urows:
            urow.StreetNum = streetnum
            urow.StreetName = streetname
            urows.updateRow(urow)

        obj_id += 1

    print 'Finished Updating Rows in ' + address_fgd

    del urow, urows, srow, srows
```

A `for` loop is implemented to cycle through the rows (in this case a single row) of the update cursor. The program only needs to update a single record at a time, so the above structure provides this ability. Without accessing a single record, all of the update rows would continue to be overwritten with each iteration of the loop for `urows` and end up with the last value of the search cursor updating all of the records in the update table.

Within the `urows` loop the *StreetNum* and the *StreetName* fields in the update table are updated with the values from the search cursor. The reader might find it helpful to look back at the variables that are set from the search cursor. (Technically, the variables for the street name and street number are not required, but are used to make the code a little easier to read and follow). To actually set the updated values in the table for the given row, the following line is used.

```
urows.updateRow(urow)
```

To ensure that the next update row is queried properly, the `obj_id` variable is incremented by one. Notice also that the `obj_id` variable is not indented as part of the `urows` loop, but is indented as part of the `srows` loop. The code must update the `obj_id` variable before it is used again in the update cursor line. Since the code is only querying a single record, the code developer can actually increment the `obj_id` value within either the `urows` or the `srows` loop. Generally it is more appropriate to increment the obj_id as part of the `srows` loop because the initial process of the code is to iterate over each record of the search cursor, not update all of the rows at the same time for the update cursor.

As shown above, the cursors are deleted to free up space and to assist with removing potential data locks.

This section illustrated the common implementations of the search, insert, and update cursors. These are valuable structures for being able to iteratively read through, create, and update records in database tables, feature classes, and individual data file formats supported by ArcGIS. Being able to effectively use cursors can assist the code builder in developing automated processes where data needs to be extracted from one database and put into another. Many of these types of data commonly require tens, to hundreds, to millions of records. These types of data and processes are good candidates for automation and can run during off hours (at night or non-peak hours). The examples and exercises provided have been building on the concept of process automation and will continue to be developed in the subsequent chapters. Chapter 11 focuses on script automation.

Table Joins

A common need for many GIS users is bring together data from a variety of sources. In some cases the different datasets are maintained separately from one another. For example a biologist may collect and maintain spatial data and some attribute characteristics for certain species of animals. Part of the information the biologist may need is data related to water temperature of nearby lakes which is maintained by a water agency. The water agency has an identification field in the lakes table called *LakeID*. When the biologist created the species feature class (e.g. a point feature class of species) one of the attributes she added was the *LakeID*. Since the biologist is not the maintainer of the water data, there is no real need to maintain lake related information as part of the animal species data; however, the biologist is interested in the lake data because it contains useful information for the biological analysis the biologist performs. The two datasets can be related to one another through a "dynamic relationship" by "virtually connecting" them via a database mechanism called a table join. Table joins allow different datasets that are separately maintained to connect to one another so that values from the respective tables can be used in attribute queries, reading values from the respective tables, updating data, or summarizing information from the "joined" tables.

Two forms of these "virtual relationships" exist in ArcGIS, 1) table joins and 2) table relates. Both types of relationships require a common attribute to exist in both datasets. A table join is most often used when a one-to-one relationship exists between the two data sets. A table relate provides the ability to perform more complex relationships between data sets where a one-to-many or many-to-many relationship exists. For example, a parcel many have many addresses within it (one parcel, many addresses) or an apartment complex many have multiple buildings that cross parcel lines and also have many addresses per building (many buildings to many parcels).

This chapter will focus only on table joins since the fundamentals are similar for both types of relationships. The reader can find more information on joins and relates in the **ArcGIS Help** under **Data Management—Managing geodatabases—Geographic Data Types—Tables**. Readers are recommended to review all of the ArcGIS Help topics under Tables. This help section covers tables in general, table design, joining, relating, as well as a special type of join, the "spatial join." Specific information about the Spatial Join routine can be found in the **ArcGIS Help** under **Analysis—Spatial Join**, since Spatial Join is a geoprocessing tool found within the Analysis toolbox. A spatial join combines attributes from different datasets that have overlapping geographies. For example each address in an address point file can be assigned the parcel number the addresses fall within by using a spatial join geoprocessing routine.

Before developing code or determining if a join function should be used the analyst should review each data source to determine what the attributes are and if there are any common attributes between the two that might be used to join them. The following shows the attribute tables from the **Lake** feature class and the **Lake_Info** table.

OBJECTID	Shape	AREA	PERIMETER	HYDRO24CA_	HYDRO24CA1	TYPE	HNAME	Shap
3	Polygon	336207	2556.76	4348	4	L	BASS LAKE	8
5	Polygon	12105.6	470.1	4365	66	L	BASS LAKE	1
1	Polygon	45009600	141633	4064	74	L	FOLSOM LAKE	46
2	Polygon	41307	1003.27	4280	5	R	HINKLE RESERVOIR	3
4	Polygon	1960410	21568.8	4352	109	L	LAKE NATOMA	70
6	Polygon	35517.3	896.588	4370	30	R	RESERVOIR	
7	Polygon	33562.4	730.349	4406	8	R	WILLOW HILL RESERVOIR	2

Lake feature class attribute table.

OBJECTID	LakeID	Lake_Name	Temp_F
1	74	Folsom Lake	65
2	109	Lake Natoma	72
3	5	Hinkle Reservoir	73
4	66	Bass Lake	63
5	8	Willow Hill Reservoir	75

Lake_Info attribute table.

Upon reviewing the two data sets one sees that a lake temperature value only exists in the **Lake_Info** attribute table and not in the **Lake** feature class. Also, the two attribute tables have different numbers of records (the **Lake** feature class has seven records; the **Lake_Info** has five records). One will also notice that Bass Lake is represented by two separate features in the **Lake** feature class; whereas, the **Lake_Info** table only contains a single temperature value for that feature. In addition, the analyst sees that a common field exists in both attribute tables, although the name of that field is different. In the **Lake** feature class the field is called *HYDRO24CA1* whereas, in the **Lake_Info** table, the attribute with the same values is called *LakeID*. These two attributes also have the same data type (a number value). These respective attributes will be used to perform a table join.

In ArcMap, a table join can easily be performed by right clicking on the feature class or table in the Table of Contents and then selecting **Joins and Relates— Join**.

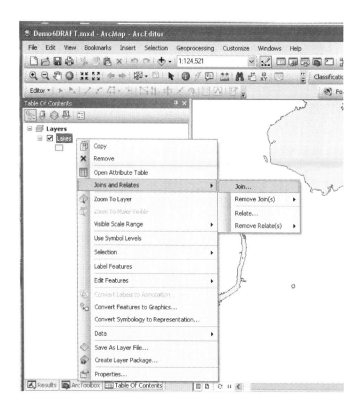

A dialog box then appears that allows the user to select the matching attributes fields from each attribute table to perform the join. The screen shot below shows the *HYDRO24CA1* attribute from the **Lakes** feature class and the *LakeID* from the **Lake_Info** table.

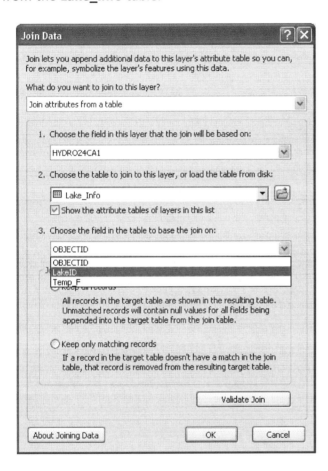

Notice that "Keep all records" is checked. This is the default. However, some users may not want to see the unmatched records during a join because they are most likely irrelevant.

Attribute Indexes

When the user clicks OK, a message appears asking if the user wants to create an attribute index. An index is a database construct that allows quick access to data values. One can think of an attribute index as an index in a book. Key topics in a book index are sorted in alphabetical order. A reader can consult the index, find the topic and locate the page number the topic is on, turn to the page in the book and begin reading about that topic. In the same manner, an attribute index creates a sorted set of values on one or more attributes and usually assigns a number to the sorted list. The index is sometimes referred to as a look up table. The user never "sees" the look up table, but it essentially contains a number and the sorted attribute value. When a "join" is performed, the index serves as a more efficient way to "connect" values from different tables.

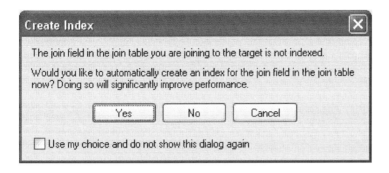

Recommendation: It is highly recommended that attribute indexes are created and used when "joining" one or more tables, especially those with large numbers of records. Indexes will need to be created for each attribute that an analyst expects to use when performing table joins and queries that involve joined tables.

For example, a city can have a master address database with hundreds of thousands of addresses. These addresses often relate to parcel or street centerline data of which each of these datasets can have hundreds of thousands of records. Attempting to maintain updates to each of these datasets and relationships between them can be very time consuming if attribute indexes are not used in the process.

In this example, after the user clicks "Yes" from above, the attribute tables from the different datasets are "joined." The user can open the attribute table for the **Lake** feature class to see the "connected" attribute tables.

Joined attribute tables. The outline represents the attributes from the *Lake_Info* table. The reader may want to do the join in ArcMap to actually see the results.

The attributes outlined above show the attribute columns from the **Lake_Info** table.

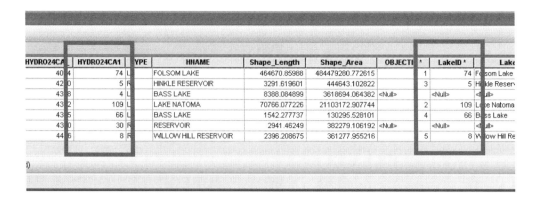

Notice that a record in the **Lake_Info** table matches records in the **Lakes** feature class where *HYDRO24CA1* and *LakeID* are the same. Note the *<null>* values in the *LakeID* attributes. The *<null>* indicates that no match was found. These show up in the resulting joined table because the user chose to show all records in the join options versus only matching records.

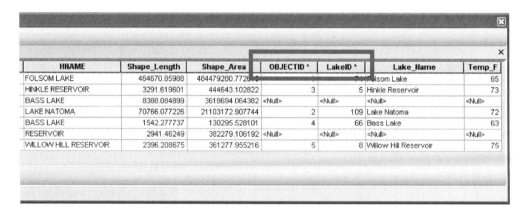

Notice that the *OBJECTID* and the *LakeID* attributes each have an asterisk (*) next to the attribute name. The asterisk indicates that an index exists for this attribute. The *OBJECTID* in all tables and feature classes are created automatically by ArcGIS.

At this point the user can create attribute queries, such as with the Select Layer By Attribute routine using the fields from the **Lake** feature class and the joined attribute table from the **Lake_Info** table. For example the attribute query "selects" records where the Lake *HYDRO24CA1* attribute is equal to the **Lake_Info** *LakeID*.

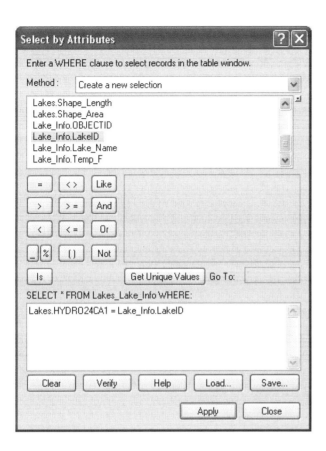

The analyst should notice that with "joined" attributes, the query looks different. With joined tables, queries must include the feature class or table name in addition to the attribute. The above query statement looks like this.

```
Lakes.HYDRO24CA1 = Lake_Info.LakeID
```

From a programming point of view, this is important to know because besides joining tables, often a subsequent step is to query the data to select features or rows or to use a query in a cursor routine. The query syntax may be a little more challenging with joined data sets and often require troubleshooting by the code developer.

Programming and Using Table Joins

To use joined tables in Python scripts, three elements need to be created:

1. Attribute Index – (not required, but good for large datasets)
2. Feature Layer or Table View
3. Table Join

The attribute index is created to efficiently perform the table join. The index is not required, but is beneficial with large datasets. The feature layer or table view is required for the table join to occur and so it can be used in feature selections by attributes or location or in cursors. Table joins cannot use "feature classes" or "tables." Feature layers and table views can be created at any time before the `AddJoin` routine is used to create the table join. The author tends to create the indexes first and then then make the feature layer or table view. See the ArcGIS Help for **Add Join**. The table join needs to be identified so the script and other processes can use and perform a join.

Create an Attribute Index

The first step to creating and using a table join is to set up an attribute index on the attribute that is expected to be used in the join. The Add Index routine is used to create the attribute index. See **Data Management Tools—Indexes—Add Attribute Index** for more details.

The required parameters for the Add Attribute Index are:

1. Table name
2. Field(s) participating in the index
3. Index name (optional, but may be helpful in scripting routines)

In the example above, the add index syntax may look like this:

```
# create indexes for the feature classes and tables
arcpy.AddIndex_management(lake_fc, 'HYDRO24CA1', 'Hydro_Index', 'NON_UNIQUE', 'NON_ASCENDING')
arcpy.AddIndex_management(lake_table, 'LakeID', 'Lake_Index', 'NON_UNIQUE', 'NON_ASCENDING')
```

Two indexes are shown above, one for a feature class, the other for a table. The first parameter is the feature class or table name (shown as variable in the screenshot). The second parameter represents the attribute the index will be created on. The third parameter is a name of the index which is any name the developer chooses. The last two parameters indicate if the index will represent unique values and if the index will be ascending or descending. Most of the time, the parameters shown above can be used.

Create Feature Layers or Table Views

Feature layers and table views can be created as described in Chapter 5 and need to be created before the `AddJoin` routine can be implemented. The screen shot below shows how a feature layer and table view are created before being used in the Add Join routine.

```
# create indexes for the feature classes and tables
arcpy.AddIndex_management(lake_fc, 'HYDRO24CA1', 'Hydro_Index', 'NON_UNIQUE', 'NON_
arcpy.AddIndex_management(lake_table, 'LakeID', 'Lake_Index', 'NON_UNIQUE', 'NON_ASC

# Create feature a feature layer and table view
# so the AddJoin routine can run
if arcpy.Exists(lake_fl):
    arcpy.Delete_management(lake_fl)

if arcpy.Exists(lake_tv):
    arcpy.Delete_management(lake_tv)

# Make Feature Layer
arcpy.MakeFeatureLayer_management(lake_fc, lake_fl)
# Make Table View
arcpy.MakeTableView_management(lake_table, lake_tv)

# Create the Join
arcpy.AddJoin_management(lake_fl, "HYDRO24CA1", lake_tv, "LakeID", "KEEP_ALL")
```

Create the Table Join

Once these elements are created, they can be used appropriately in the `AddJoin` routine. In addition to the feature layer and table view, the other required parameters are the respective "join fields" that is used to join the two data sets together. In the example, the **Lake** feature layer uses the *HYDRO24CA1* field and the **Lake_Info** table uses the *LakeID* field. Note that the `AddJoin` routine uses the "KEEP_ALL" value to keep all of the records from the input feature layer, even if there is no match in the **Lake_Info** table. Unmatched records are not "deleted" or "eliminated" since a join is performed in memory and does not physically change the data.

Using and Accessing Information in Joined Data

Once the join is complete, the values from both attribute tables can be used for subsequent processing. The figure below shows the `ListFields` routine to print out a list of fields to the Python Shell as well as set up a search cursor to read specific elements from the joined data.

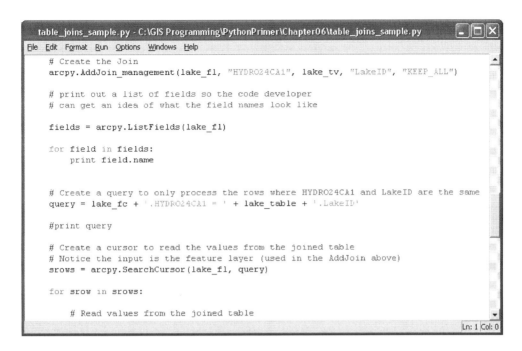

The `ListFields` routine(and using lists in general) will be discussed in the next chapter. This routine may be useful for a code developer to see a list of fields, since joined fields look different and require some syntax changes when developing queries (such as selecting data or creating cursors). The program continues to read values from both the feature class and the table as shown below. These are printed to the Python Shell, but they could easily have been written to a separate table or used to update values in another feature class, table or in other subsequent processes within the code.

```python
    # Create a cursor to read the values from the joined table
    # Notice the input is the feature layer (used in the AddJoin above)
    srows = arcpy.SearchCursor(lake_fl, query)

    for srow in srows:

        # Read values from the joined table
        # Note the feature class name and the table name are used
        HydroID = srow.getValue(lake_fc + '.HYDRO24CA1')
        LakeName = srow.getValue(lake_fc + '.HNAME')
        LakeID = srow.getValue(lake_table + '.LakeID')
        temp_F = srow.getValue(lake_table + '.Temp_F')

        # print the values out to the Python Shell
        # these values can be used in update cursors for
        # populating rows in other tables or feature classes

        print lake_fc + ' HYDRO24CA1 is: ' + str(HydroID)
        print lake_fc + ' Lake Name is: ' + LakeName
        print lake_table + ' LakeID is: ' + str(LakeID)
        print lake_table + ' Temperature is: ' + str(temp_F) + '\n'

    # remove the join, since it is no longer used
    arcpy.RemoveJoin_management(lake_fl, lake_table)

    # delete cursors to free up memory and help elminate data locks
    del srow, srows

except:
```

Notice in the above search cursor that the values obtained from the joined feature class and table have the feature class or tables' root name (using a variable) and the respective field name using the following general syntax

```
<feature class or table name>.<field_name>
```

The feature class or the table name must be included so that the Python script knows which field is associated with it. The above example uses variables that reference a feature class or table within a file geodatabase, so the variable can be used directly in the getValue parameter. If the source feature class or table references a shapefile (e.g. MyFeatureClass.shp) or dBase table (e.g. MyTable.dbf), then either the "hard coded" root name of the feature class or table must be used or a variable representing the root name must be used. For example,

```
aFeatureClass = 'MyFeatureClass.shp'
```

```
aTable = 'MyTable.dbf'

aFeatureClass_root = 'MyFeatureClass'
aTable_root = 'MyTable'
```

NOTE: Other Python syntax can be used to strip off the root name.

In addition, note the use of the `.getValue` method. This method must be used, since a string value is being represented by both a variable and a specific attribute name. The other method to obtain the value of a column

```
srow.<name of field>
```

will not work since a variable and a string cannot be combined to form the proper syntax for Python to understand.

Summary

As has been shown in this chapter a variety of cursor methods and the table join operations have been illustrated to show how code developers can access specific values from feature classes or tables (even joined tables) and use those values for other purposes. These operations can be very useful when working with large databases, performing frequent updates, and summarizing or combining data from different databases or systems. Often these structures are found in automated scripts that run during off-hours or off-peak computer system loads. Properly constructing both the cursor and the processing logic to successfully process the records are key concepts the reader should take away from this chapter. In many cases, designing the proper processing logic will continue to be a primary challenge when using cursors. Once this is overcome, a Python script can become very useful to an organization managing large datasets and databases.

Demos Chapter 6

The demos for this chapter are broken into four parts:

1. Demo 6a – Search Cursor
2. Demo 6b – Insert Cursor
3. Demo 6c - Search and Update Cursor
4. Demo 6d – Table Joins

The demos follow the procedures mentioned above. The data for these demos can be found in **\PythonPrimer\Chapter06\Data**. The reader may find it helpful to view the feature classes and tables in ArcCatalog and can also review data in ArcMap.

The concepts illustrated in these demos are:

ArcGIS Concepts

Search Cursor
Insert Cursor
Update Cursor
Create Tables
Add Fields
Data paths
Queries
Data locks
Get and Set values

Python Concepts

`for` **loop**
`while` **loop**
Indentation
Casting numbers to strings
`os` **module**

Demo 6a: Search Cursor

This demo uses the **Sacramento_Streets.shp** file that can be found in the **\PythonPrimer\Chapter06\Data** folder. Open ArcMap or ArcCatalog to see the various attributes and values of **Sacramento_Street.shp** file as needed.

NOTE: The reader's path may be different than shown below and depending on where the data is put.

This demonstration illustrates the construction and use of a search cursor to read values from an attribute table using a query to read only the highway records.

1. Add the `arcpy`, `sys`, and `traceback` modules. Recycle the `except:` block from a previous script. Add some commentary describing the script.

2. Set a workspace to the data folder shown above.
3. Create a variable to hold the value of the shapefile

At this point, the script should look similar to this.

```
# Demo 6a - Search Cursor Example
# Created by:  Nathan Jennings
#              www.jenningsplanet.com
# Created on: 01.29.2011
# Updated on: 10.30.2011
# Copyright:  2011

'''
This demo illustrates the use of the search cursor types used in ArcGIS.

Users may need to change the workspace or data paths
'''

import arcpy, sys, os, traceback

# set the current workspace (in the case a folder)
arcpy.env.workspace = 'C:\\PythonPrimer\\Chapter06\\Data\\'
# assign a variable to hold the street shapefile
streets_shp = 'Sacramento_Streets.shp'


try:
```

Since only highways will be used in this cursor, a query variable is set with the following query:

4. Create a query variable

    ```
    query = '"CLASS" = \'H\''
    ```

5. Create the search cursor using the `streets_shp` variable and the query variable.

```
Demo6a.py - C:\PythonPrimer\Chapter06\Demo6a.py
File  Edit  Format  Run  Options  Windows  Help

import arcpy, sys, os, traceback

# set the current workspace (in the case a folder)
arcpy.env.workspace = 'C:\\PythonPrimer\\Chapter06\\Data\\'
# assign a variable to hold the street shapefile
streets_shp = 'Sacramento_Streets.shp'


try:

    # set a variable to hold the query string
    query = '"CLASS" = \'H\''

    # Create a search cursor to access rows that
    # have a road class of "H" - Highway

    # Gets a collection of rows from a feature class or table
    srows = arcpy.SearchCursor(streets_shp, query)

Ln: 12 Col: 0
```

Now that the cursor is defined, a looping structure can be created to iterate over each of the records in the cursor.

6. Create a `for` loop to iterate over the rows of the cursor

Since a search cursor simply reads values from a table, create two new variables to hold the values of the following attributes:

FULLSTREET
UNIQUE_ID

7. Create two variables for the attributes above. Remember, two different methods can be used to obtain the values for a specific row (record) and attribute (column).

```
try:

    # set a variable to hold the query string
    query = '"CLASS" = \'H\''

    # Create a search cursor to access rows that
    # have a road class of "H" - Highway

    # Gets a collection of rows from a feature class or table
    srows = arcpy.SearchCursor(streets_shp, query)

    for srow in srows:    # for each row in the cursor

        # assign a variable for the value of srow.FULLSTREET
        # assign a variable for the value of srow.getValue("UNIQUE_ID")
        fullstreet =  srow.FULLSTREET
        unique_id = srow.getValue("UNIQUE_ID")
```

For this demo, the code developer simply prints the values to the Python Shell.

8. Create a print statement that prints the full street name and the `unique_id` value. Remember that when printing numbers with a print string, that the `str()` must contain the value.

```
    # Gets a collection of rows from a feature class or table
    srows = arcpy.SearchCursor(streets_shp, query)

    for srow in srows:    # for each row in the cursor

        # assign a variable for the value of srow.FULLSTREET
        # assign a variable for the value of srow.getValue("UNIQUE_ID")
        fullstreet =  srow.FULLSTREET
        unique_id = srow.getValue("UNIQUE_ID")

        # prints the value of the variables
        # i.e. the value of srow.FULLSTREET and srow.GetValue("UNIQUE_ID")
        print fullstreet + ' ' + str(unique_id)

except:
```

9. Save and check the module. Fix any problems encountered.
10. Save and run the script. The Python Shell should display the values read from the **Sacramento_Streets.shp** file. Only highways should appear in the printed statements.

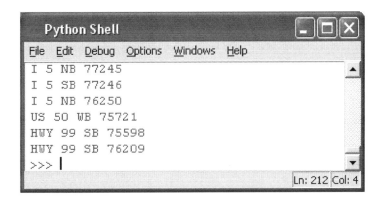

Demo 6b: Insert Cursor

This demonstration will illustrate the use of the insert cursor. In addition, the user will learn how to create a table from scratch and add some fields. The insert cursor will be used to create new rows and populate initial values to one of the attributes.

This demo uses a pre-existing file geodatabase (**cursors.gdb**) to create the **addresses** table. The geodatabase can be found in the **\PythonPrimer\Chapter06\Data** folder.

NOTE: The reader's path may be different than shown below depending on where the data exists.

1. Start the script by adding the `arcpy`, `sys`, and `traceback` modules. Add some commentary describing the script. Recycle the `except:` block from a previous script.

2. Add a workspace (`arcpy.env.workspace`) and a workspace variable (`outpath`) to point to the file geodatabase. Both of these will point to the full path to the **cursors.gdb** file geodatabase. Note: a workspace to a file geodatabase includes the folder plus the name of the file geodatabase with the extension (.gdb).

3. Add a variable (`address_table`) to hold the name of the new table called **addresses**. There is no file extension on the table, since it resides within the file geodatabase. Tables and feature classes within file, personal, or SDE geodatabases do not have extensions.

The code should look similar to the following:

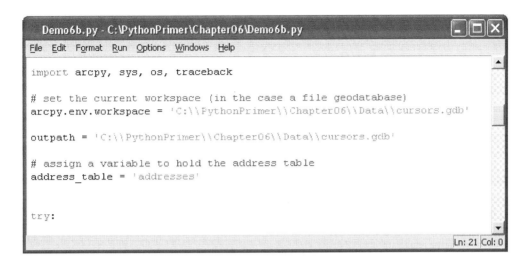

Since this demo illustrates creating a new table and adding attributes, the `CreateTable` function must be used.

4. Add the `CreateTable` routine to the script.

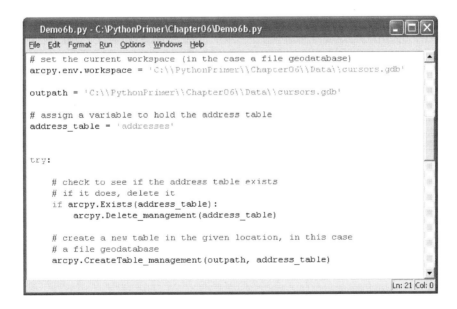

5. The `CreateTable` routine requires an output path and a table name. Use the `output_path` (which is the same as the workspace) and the `address_table` variable for the table name. Use these two variables in the `CreateTable` routine.

Since the script may be run multiple times, it is good practice that if the developer knows a table or feature class will be replaced with a new feature class with the same name, the use of the `Exists` function can be used to test to see if the table already exists. If it does, the `Delete` routine can be used to delete the table. If it is not deleted, the code will break, since it will be trying to create a table with the same name of the table that already exists.

6. Add the `Exists` and `Delete` statements above. Notice that an `if` statement is used to check to see if the table exists.

Now that an empty table has been created, some attributes (fields) are required before rows are added. The `AddField` ArcGIS routine is used to perform this operation. Consult the ArcGIS Help for `AddField` to find out more details on using different data types and how to write the syntax. Most of the time in ArcGIS, the fields will be strings with a specified length, integers (long or short), or doubles (floating point numbers) values.

7. Use the `AddField` routine to add the following fields with the specified data type. Each added field uses the same general syntax.

AddID - Long
StreetNum - Long
StreetName – Text, length 60 characters

```
try:

    # check to see if the address table exists
    # if it does, delete it
    if arcpy.Exists(address_table):
        arcpy.Delete_management(address_table)

    # create a new table in the given location, in this case
    # a file geodatabase
    arcpy.CreateTable_management(outpath, address_table)

    # add fields to the table
    arcpy.AddField_management(address_table, 'AddID', 'LONG')
    arcpy.AddField_management(address_table, 'StreetNum', 'LONG')
    arcpy.AddField_management(address_table, 'StreetName', 'TEXT', '', '60')

    print 'Created table and added fields'
```

Note that the `StreetName` text field show that it is defined as 'TEXT' with a length of 60 characters and that an "empty" parameter is shown with (''). The single quotes shown above represent a placeholder for an optional parameter that is not used for the geoprocessing routine. When optional parameters are not needed and do not occur at the end of the geoprocessing routine, a place holder using either a pair of single or double quotes is required. The reader should pay close attention when reviewing ArcGIS routines that contain optional parameters. The order of the parameters must be maintained, even if some of them are not used within the geoprocessing routine.

Now that a blank table has been created and has some attribute fields, the insert cursor can be implemented. If a pre-existing table contained attribute fields, the above steps are not needed.

8. Add the insert cursor routine. The only parameter required is the name of the table to insert records to. In this case, **addresses** table (i.e. the variable `address_table` is used in the insert cursor).

```
try:

    # check to see if the address table exists
    # if it does, delete it
    if arcpy.Exists(address_table):
        arcpy.Delete_management(address_table)

    # create a new table in the given location, in this case
    # a file geodatabase
    arcpy.CreateTable_management(outpath, address_table)

    # add fields to the table
    arcpy.AddField_management(address_table, 'AddID', 'LONG')
    arcpy.AddField_management(address_table, 'StreetNum', 'LONG')
    arcpy.AddField_management(address_table, 'StreetName', 'TEXT', '', '60')

    print 'Created table and added fields'

    # Create cursor object
    irows = arcpy.InsertCursor(address_table)
```

Once the insert cursor is created, a looping structure is required to add the respective number of rows. A `while` loop is suitable for this process. A `counter` variable can be used to initialize the loop and establish a certain number of records (rows) in the table. NOTE: A `for` loop could also have been used.

9. Add the `while` looping structure and `counter` variable.

```
# create a new table in the given location, in this case
# a file geodatabase
arcpy.CreateTable_management(outpath, address_table)

# add fields to the table
arcpy.AddField_management(address_table, 'AddID', 'LONG')
arcpy.AddField_management(address_table, 'StreetNum', 'LONG')
arcpy.AddField_management(address_table, 'StreetName', 'TEXT', '', '60')

print 'Created table and added fields'

# Create cursor object
irows = arcpy.InsertCursor(address_table)

# counter variable to initialize loop
counter = 1

# run the loop until counter is <= to 10
while counter <= 10:
```

Within the loop, the actual new rows are created and initialized. Essentially three parts are needed.

    a. Create a new row (using the `newRow` method)
    b. Initialize attribute values (as needed)
    c. Insert the new row using the cursor (using the `insertRow` method)

10. Add the elements listed above. Notice how the attribute `AddID` is used when it is initialized (assigned) with the value of the `counter` variable. For example, when `counter` equals 1, a specific row in the attribute table the attribute `AddID` will equal 1; when `counter` is incremented to the value 2; `AddID` for a given row will equal 2.

```
arcpy.AddField_management(address_table, 'StreetName', 'TEXT', '', '60')

print 'Created table and added fields'

# Create cursor object
irows = arcpy.InsertCursor(address_table)

# counter variable to initialize loop
counter = 1

# run the loop until counter is <= to 10
while counter <= 10:

    irow = irows.newRow()      # create a new row
    irow.AddID = counter       # update the AddID field
                               # by calculating it to
                               # the counter value

    irows.insertRow(irow)      # actually inserts the row into the table
    counter += 1               # increment the counter so a a new row can be added
```

The final part of the Demo6b script includes the `del` (delete) Python function to remove the cursor variables from memory which can also help to eliminate data locking problems. A `print` statement is also added to indicate the number of rows added.

```
irows = arcpy.InsertCursor(address_table)

# counter variable to initialize loop
counter = 1

# run the loop until counter is <= to 10
while counter <= 10:

    irow = irows.newRow()      # create a new row
    irow.AddID = counter       # update the AddID field
                               # by calculating it to
                               # the counter value

    irows.insertRow(irow)      # actually inserts the row into the table
    counter += 1               # increment the counter so a a new row can be added

    del irow, irows    # delete the cursor varaible to prevent schema locks

    print 'Added ' + str(counter - 1) + ' rows to table: ' + address_table

except:
```

When the script is executed and completes, 10 rows are added to the address table that was created from scratch. Notice how the `while` loop is structured and the primary components of creating and initializing rows are used in addition to incrementing the `counter` variable by one each time the `while` loop executes. Note also that the `print` statement uses "counter - 1" instead of only `counter`. The reason for this is that the `counter` variable is used in the `while` statement to check to see if `counter` is greater than 10. If it is, then the loop stops. At the last iteration of the `while` loop, `counter` is 11. To report out the correct number of rows added, `counter` must have 1 subtracted from it when used in the `print` statement.

The resulting table should look like the following illustration after the insert cursor script is completed. Use ArcCatalog to view the resulting table. Make sure to close the Python IDLE and Shell windows before looking at the table in ArcCatalog or use View—Refresh (or F5) to refresh the address table within the cursors.gdb file geodatabase.

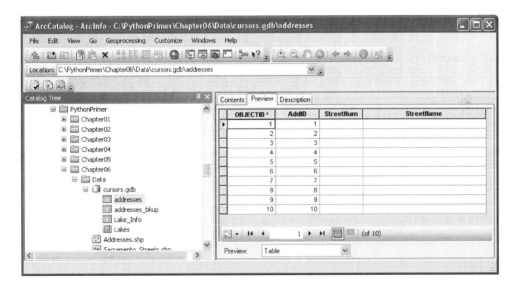

Demo 6c: Search and Update Cursor

This demo uses the **addresses.shp** file that can be found in the **\PythonPrimer\Chapter06\Data** folder. Open ArcMap or ArcCatalog to see the various attributes and values of **addresses.shp** file as needed. In addition, a file geodatabase (**cursors.gdb**) is provided that contains the **addresses** table from the insert cursor in **Demo 6b**. If the **addresses** table does not exist, follow the directions in **Demo 6b** to create it.

This demonstration illustrates the construction and use of a search cursor that contains the query ("where clause") parameter to obtain the first 10 records from the **addresses.shp** file attribute table. For each of the 10 records the *STREETNUMB* and *STREETNAME* values are "read" using the search cursor and then are used with an update cursor to "update" the *StreetNum* and *StreetName* attributes in the **addresses** table in the file geodatabase.

1. Add the `arcpy`, `sys`, and `traceback` modules. Recycle the `except:` block from a previous script. Add some commentary describing the script.

2. Set a workspace to the data folder shown below for the shapefile and create a separate variable to hold the data path the file geodatabase (**cursors.gdb**).

NOTE: The reader's path may be different than shown below and depending on where the data is located.

```
# Created by:  Nathan Jennings
#              www.jenningsplanet.com
# Created on: 01.29.2011
# Updated on: 10.30.2011
# Copyright:  2011

'''
This demo illustrates the use of the search and update cursor types
used in ArcGIS.  This script reads from one data file (addresses.shp) and
write values to the file geodatabase table (addresses).  Note the file
geodatabase contains the table, not a feature class.
'''

import arcpy, sys, os, traceback

# set the current workspace (in the case a folder)
arcpy.env.workspace = 'C:\\PythonPrimer\\Chapter06\\Data\\'

fgdpath = 'C:\\PythonPrimer\\Chapter06\\Data\\cursors.gdb\\'

address_shp = 'addresses.shp'
address_fgd = fgdpath + 'addresses'


try:
```

As indicated in the chapter, note the "workspace" versus the variable that holds the data path to the file geodatabase. Only one workspace can be set at a time. Notice that the workspace is set to the path where the shapefile is located. Also note the variable used to store the name of the shapefile versus the variable to store both the path and the name of the address table in the file geodatabase. The address_fgd represents a string that points to the entire path including the name of the file geodatabase name. Inside of the geodatabase, the **addresses** table exists. Because a workspace is used to point to the path where the shapefile is stored, only a variable for the **addresses.shp** is needed.

Because no new feature classes or tables are being created, a workspace really is not needed and the code developer could have created a variable to hold just the path (or the path and file name) for the shapefile.

3. Next a query variable is created, since in this program it is desired to query only the first ten records of the shapefile. The query will use the *FID* values from the **addresses.shp** file.

   ```
   query = '"FID" >= 0 AND "FID" <= 10'
   ```

4. Create the search cursor using the query and the variable for the address shapefile (`address_shp`) for the input data set.

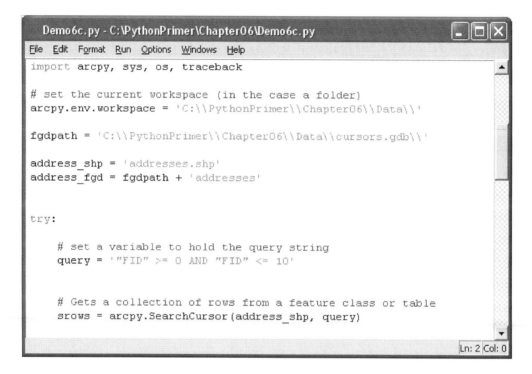

5. Create a `for` loop structure to iterate over the records (rows or features) in the search cursor.

6. Create variables to hold the values of the *STREETNAME* and *STREENUMB* attributes for the specific row in the cursor. Use one or both methods shown below to "get" or "read" a value from the respective attribute. The `obj_id` variable is used in the next step.

```
Demo6c.py - C:\PythonPrimer\Chapter06\Demo6c.py
File  Edit  Format  Run  Options  Windows  Help

arcpy.env.workspace = 'C:\\PythonPrimer\\Chapter06\\Data\\'

fgdpath = 'C:\\PythonPrimer\\Chapter06\\Data\\cursors.gdb\\'

address_shp = 'addresses.shp'
address_fgd = fgdpath + 'addresses'


try:

    # set a variable to hold the query string
    query = '"FID" >= 0 AND "FID" <= 10'


    # Gets a collection of rows from a feature class or table
    srows = arcpy.SearchCursor(address_shp, query)

    obj_id = 1

    for srow in srows:

        # assign a for the value of srow.STREETNAME
        # assign a variable for the value of srow.getValue("STREETNUMB")
        streetname = srow.STREETNAME
        streetnum = srow.getValue("STREETNUMB")
```

Once the *STREETNUMB* and *STREETNAME* attribute values are obtained for a row in the **addresses.shp** file, an update cursor can be constructed so that a row in the **addresses** file geodatabase table can be accessed to update the *StreetNum* and *StreetName* attributes. Only a single row needs to be accessed in the **addresses** table. A query can be constructed to only access the proper row by using the *OBJECTID* field from the **addresses** table. In this example it is assumed that the ObjectIDs (*FID*) in the **addresses.shp** file match the Object IDs (*OBJECTID*) of the **addresses** file geodatabase table.

7. Set up the query for the update cursor.

To create the proper syntax for the query a variable is created and assigned a value of 1 (e.g. `obj_id = 1`). This variable is placed just before the `for` loop that iterates over the search cursor rows. (Actually, the `obj_id` variable can be defined anywhere before the update cursor is defined).

Since the idea for the update cursor is to update only a single row in the **addresses** table in the file geodatabase, a specific row must be accessed. To do this a query must be set up to access the specific row (in this case the appropriate *OBJECTID* attribute value for a respective row). A variable can be used to dynamically access the given row. In this example the first row accessed is where the OBJECTID = 1 (the first row of the **addressees** file geodatabase table). Using a variable to store a value that can change with each iteration of the looping structure is desirable, since without a query to limit the update cursor to access a single row, the update cursor would iterate through all of the rows and update each value with the same value.

Notice the input data set for the update cursor is the **addresses** file geodatabase table and not the **addresses.shp** file feature class.

NOTE: The user may want to comment out the query statement and not use a query parameter in the update cursor to see what effect the query has.

```python
# set a variable to hold the query string
query = '"FID" >= 0 AND "FID" <= 10'


# Gets a collection of rows from a feature class or table
srows = arcpy.SearchCursor(address_shp, query)

obj_id = 1

for srow in srows:

    # assign a for the value of srow.STREETNAME
    # assign a variable for the value of srow.getValue("STREETNUMB")
    streetname = srow.STREETNAME
    streetnum = srow.getValue("STREETNUMB")

    # Get a collection of rows for the update cursor
    # based on a query
    # This update cursor will only return a single
    # record, which is desired.
    query = '"OBJECTID" = ' + str(obj_id)

    # create the update cursor
    urows = arcpy.UpdateCursor(address_fgd, query)
```

8. Create a `for` loop structure for the update cursor

A `for` loop is used to iterate through the rows of the update cursor (in this case only a single row).

9. Use the row method to update the respective update attribute with the value from the search cursor. For this example, the format `urow.StreetNum` or `urow.StreetName` is used. Notice that the variables `streetnum` and `streetname` store the values from the search cursor. This syntax is actually assigning the attribute values in the "update" attribute table (**addresses**) with the values from the "search" attribute table (**addresses.shp**)

10. The final step in the loop is to actually update the row with the `updateRow` method. This method actually "writes" the changes back to disk where the feature class or table is stored (in this case, the **addresses** file geodatabase table).

```
    for srow in srows:

        # assign a for the value of srow.STREETNAME
        # assign a variable for the value of srow.getValue("STREETNUMB")
        streetname = srow.STREETNAME
        streetnum = srow.getValue("STREETNUMB")

        # Get a collection of rows for the update cursor
        # based on a query
        # This update cursor will only return a single
        # record, which is desired.
        query = '"OBJECTID" = ' + str(obj_id)

        # create the update cursor
        urows = arcpy.UpdateCursor(address_fgd, query)

        # cycle through the rows (in this case only 1)
        # to actually update the row in the cursor
        # with the values obtained from the search cursor
        for urow in urows:
            urow.StreetNum = streetnum
            urow.StreetName = streetname
            urows.updateRow(urow)

        obj_id += 1

    print 'Finished Updating Rows in ' + address_fgd

    del urow, urows, srow, srows
except:
```

11. Note that the variable `obj_id` is incremented by 1 and is at the same indentation level as the lines of code within the `for` loop for the search cursor. The purpose of this is to properly increment the *OBJECTID* attribute value (in the **addresses** file geodatabase table) so that the search cursor row matches that of the update cursor and the correct row in the update cursor is updated properly (i.e. the row in the file geodatabase **addresses** table matches the same row in the **addresses.shp** attribute table, where **addresses.shp** *FID* = **addresses** *OBJECTID* attribute values).

12. A final print statement is written to let the user know the process is completed and has updated the addresses table in the file geodatabase.

13. The cursors are deleted to free up space in to help eliminate data locks using the `del` Python routine.

Demo 6d: Joining Tables

Using the data described in Chapter 6, this demo will illustrate the steps needed to perform a table join between a feature class and a standalone table. NOTE: This method can also work with two feature classes or two tables, provided that table views and/or feature layers are properly constructed.

The **Demo6d.py** script can be found in the **\PythonPrimer\Chapter06** folder. The data used for this demo can be found in the **\PythonPrimer\Chapter06\Data\cursors.gdb** file geodatabase. The **Lakes** feature class and **Lake_Info** table will be used in this demo.

1. Set up a workspace and variables for the feature class, feature layer, table, and table view. See below.

NOTE: The reader's path may be different than shown below and depending on where the data is located.

```
# Created on: 02.06.2011
# Updated on: 10.30.2011
# Copyright:  2011

'''
This demo illustrates the use of attribute indexes and table joins to perform
attribute selections and queries in a search cursor.
'''

import arcpy, sys, os, traceback

# set the current workspace (in the case a folder)
arcpy.env.workspace = 'C:\\PythonPrimer\\Chapter06\\Data\\cursors.gdb\\'

#fgdpath = 'C:\\PythonPrimer\\Chapter06\\Data\\cursors.gdb\\'

lake_fc = 'Lakes'       # this is the Lakes feature class in the cursors.gdb
lake_table = 'Lake_Info'  # this is the Lake_Info table in the cursors.gdb

lake_fl = 'Lakes_FL'    # this is a string that represents the Lakes feature layer
lake_tv = 'Lakes_TV'    # this is a string that represents the Lakes_Info table view

try:
```

2. Create a list of indexes for both the feature class and the table to check if a specific index exists. The `ListIndexes` routine is used. If the index is found, then use the `RemoveIndex` routine to remove (delete) it.

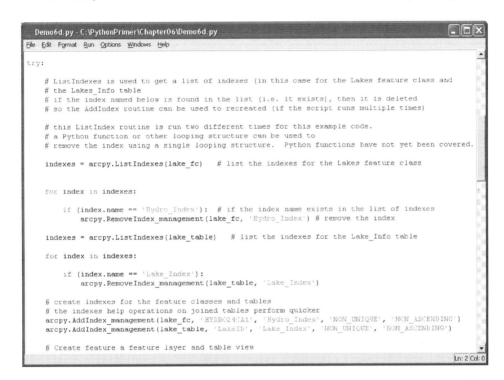

3. Once the indexes are removed (provided they previously existed), run the `AddIndex` routine to create a new index. This is performed twice, once on the **Lakes** feature class and once on the **Lake_Info** table. Refer to the documentation above or the ArcGIS Help for specific comments on the parameters. The *HYDRO24CA1* is the field in the feature class which will have an index; the name of the index is *'Hydro_Index'* which is just a string. The *LakeID* is the field in the table that will have an index; the name of the index is *'Lake_Index'*. See the above figure or refer to the Demo6d.py script. The *'NON_UNIQUE'* and the *'NON_ASCENDING'* parameters are the defaults for the `AddIndex` routine.

4. Since the join requires a feature layer and a table view (and not a feature class or a table), these must be created. The next steps show the `MakeFeatureLayer` and the `MakeTableView` being used to create

the feature layer and the table view for the **Lakes** feature class and the **Lake_Info** table, respectively.

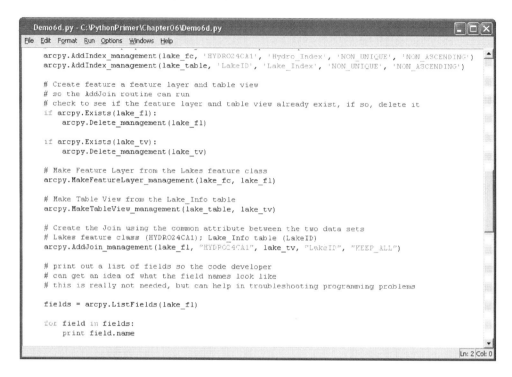

5. Once the feature layer and table view are created, the join can occur by using the `AddJoin` routine. Notice the feature layer (**Lakes**) will have the table view (**Lake_Info**) joined to it using the *HYDRO24CA1* and the *LakeID* fields. All of the records from the resulting join will be shown using the *KEEP_ALL* parameter.

6. When developing code for joining tables, it might be a good idea to list out the field names resulting from the join. This can be useful because the field names change slightly and take the form:

   ```
   <feature class or table name>.<field name>
   ```

   A `ListFields` routine can be used to access the fields from a feature class or table. To print out a list of the fields to the Python Shell a `for` loop can be used to print the field names. Lists will be discussed in Chapter 7. This step is not required for using fields and values from joined data, but it may be helpful for troubleshooting purposes. See the above figure.

7. After the data sets have been joined, specific records and values can be accessed and used. A search cursor is implemented to retrieve a set of records from the data. In this case, since some **Lake_Info** *LakeID* attribute values do not match the **Lakes** *HYDRO24CA1* attribute values, the code developer only wants to obtain records that have actual values in both the feature attribute table (**Lakes**) and the standalone table (**Lake_Info**), a query is used to limit the records returned from the search cursor.

    The query string is:

    ```
    query = lake_fc + '.HYDRO24CA1 = ' + lake_table + '.LakeID'
    ```

    Note that the `lake_fc` and `lake_table` variables are used to construct the query. The programmer could have "hard coded" the specific feature class and table names, but if the names of the feature class and table change while maintaining the same attribute names, the code remains flexible for this possibility.

```
fields = arcpy.ListFields(lake_fl)

for field in fields:
    print field.name


# Create a query to only process the rows where HYDRO24CA1 and LakeID are the same
query = lake_fc + '.HYDRO24CA1 = ' + lake_table + '.LakeID'

#print query

# Create a cursor to read the values from the joined table
# Notice the input is the feature layer (which now has the Lake_Info table joined to it)
srows = arcpy.SearchCursor(lake_fl, query)

for srow in srows:

    # Read values from the joined table
    # Note the feature class name and the table name are used in conjunction with
    # the respective variable that corresponds to the Lakes feature class or Lake table

    HydroID = srow.getValue(lake_fc + '.HYDRO24CA1')
    LakeName = srow.getValue(lake_fc + '.HNAME')
    LakeID = srow.getValue(lake_table + '.LakeID')
    temp_F = srow.getValue(lake_table + '.Temp_F')

    # print the values out to the Python Shell
    # these values can be used in update cursors for
    # populating rows in other tables or feature classes

    print lake_fc + ' HYDRO24CA1 is: ' + str(HydroID)
    print lake_fc + ' Lake Name is: ' + LakeName
    print lake_table + ' LakeID is: ' + str(LakeID)
    print lake_table + ' Temperature is: ' + str(temp_F) + '\n'

# remove the join, since it is no longer used
```

Notice above how the feature class and the lake table variables are used in the query string. This is another place that code developers are often challenged and one reason why printing out the joined field names may be useful. Listing out the field names can help the code developer write the correct syntax for the query. The query string follows the syntax shown above for joined field names. In this case, the query will retrieve records where the *HYDRO24CA1* field value from the **Lakes** feature class equals the *LakeID* field from the **Lake_Info** table.

The user can remove the comment sign (#) in front of the `print query` statement to print the query to the Python Shell. This may also help to troubleshoot code.

8. Once the records have been retrieved by a cursor, specific values can be accessed and used. In this case, the values are simply printed to the

Python Shell, but can easily be used with other cursor methods or processes.

Notice the `getValue` method is used (instead of using the actual field name) because the joined table has a (".") between the feature class or table name and the name of the field. In addition, notice how the feature class and table variables are used to retrieve the specific value from the respective field.

9. Once the records have been accessed and processed the join is no longer used in the script and can be removed by using the `RemoveJoin` routine. Notice that `RemoveJoin` is not indented as part of the `for` loop. If this were not the case, the join would be removed after the first iteration of the loop, which is not desired.

```
        HydroID = srow.getValue(lake_fc + '.HYDRO24CA1')
        LakeName = srow.getValue(lake_fc + '.HNAME')
        LakeID = srow.getValue(lake_table + '.LakeID')
        temp_F = srow.getValue(lake_table + '.Temp_F')

        # print the values out to the Python Shell
        # these values can be used in update cursors ror
        # populating rows in other tables or feature classes

        print lake_fc + ' HYDRO24CA1 is: ' + str(HydroID)
        print lake_fc + ' Lake Name is: ' + LakeName
        print lake_table + ' LakeID is: ' + str(LakeID)
        print lake_table + ' Temperature is: ' + str(temp_F) + '\n'

    # remove the join, since it is no longer used
    arcpy.RemoveJoin_management(lake_fl, lake_table)

    # delete cursors to free up memory and help elminate data locks
    del srow, srows

except:
```

The reader may notice that the `RemoveJoin` routine is using a feature layer and a table and not a table view. The reason for this is the `RemoveJoin`'s second parameter ("join name", a string) requires the name of the table, not the name of the table view while the first parameter (input feature layer or table view, requires the feature layer). In this case the **Lake_Info** table name is "Lake_Info"; whereas, the table view name is "Lakes TV". If the table was "aTable.dbf", the join name would be "aTable". See the ArcGIS Help for the `RemoveJoin` routine.

10. The cursor pointers are deleted to help eliminate data locks using the `del` routine.

Exercise 6: Using Cursors and Table Joins

Exercise 6 will bring together a number of elements already experienced in the course. Part of Exercise 6 provides the opportunity for an analyst to do some research to find some of the answers for specific tools that are required as well as the parameters that are needed. Refer to the *Python Primer* text as well as ArcGIS Tool Help to identify these tools and parameters.

Write a script with the following conditions.

a. Using **Demo6d.py** as a starting point, instead of writing the joined values to the Python Shell, write the values to a new output table within a pre-existing file geodatabase that the reader creates. The geodatabase will be created in ArcCatalog. Creating and populating the new output table with records will be done programmatically. See below.
b. Use an insert cursor method to write out the joined values to the new table. An update cursor will not be required.
c. **Extra** - In addition to the table, write the values out to a new feature class within the same script using the `CopyFeatures` routine. Do not use a cursor to do this.

NOTE: Most of the existing code will be used unmodified to perform this exercise. The search cursor values (i.e. `srow.getValue(<field>)`, those found in the `srow` for loop) will need to be changed to meet the requirements of the exercise. All of the other major steps -- indexes, join, feature layers, and query used in the search cursor will not need to be modified. Additional variables and data paths will likely be required in addition to the use of specific ArcGIS tools. These will need to be researched and the specific parameters added as required. The existing chapters from *A Python Primer* and the ArcGIS Help documentation should provide enough information to perform these tasks.

The program will use the following routines. Many of these already exist in the **Demo6d.py** script.

1. *CreateTable* – to create the new output table
2. *AddField* – to add one or more fields to the new output table

3. *ListIndexes* – to list any existing indexes in the datasets
4. *AddIndex* – to create an index on the join item for each table
5. *AddJoin* – to join the two attribute tables
6. *RemoveJoin* – to remove the join after it is no longer needed
7. *MakeFeatureLayer* – to make a feature layer from a feature class
8. *MakeTableView* – to make a table view from a table
9. *ListFields* – to list fields out
10. *SearchCursor* – to read values from the joined table
11. *InsertCursor* – to insert the rows and values into the new output table
12. *getValue* and *setValue* – to get and set values from the search and insert cursor routines
13. *CopyFeatures* – to write out all of the "joined" features to a new file geodatabase feature class
14. *Exists/Delete* – to check and delete datasets as needed
15. *del* – to delete cursors to free up data locks

Recommendation

After making the file geodatabase, copy the Chapter06 folder to a different location as a "back up", just in case data or files are deleted. Also make sure to write the Python script in a separate folder from **\PythonPrimer\Chapter06\Data** or **\PythonPrimer\Chapter06\MyData**. If a workspace (instead of a feature class or table) is deleted by accident and the workspace is a folder (such as \Chapter06\Data) any files in this folder may be deleted (any kind of file). It is good practice to write and process the script in a different location than where data is written to, especially if data is to be deleted.

Notes

The following will be done programmatically. The output table will be stored in the file geodatabase.

A current workspace is already set to the file geodatabase (**\PythonPrimer\Chapter06\Data\cursors.gdb**). Only one current workspace can be used at a time. The current workspace is used for the joining of the two data sets and using the search cursor to read values from the joined data. Since an output table will be created and written to while the search cursor is reading from the data in the current workspace, a different format is required to access a different location on the computer disk.

When creating the new table and setting up a cursor to write values to the new table you will need to use a format like this for a path and table name variables.

```
outpath = 'c:\\temp\\out_fgeodatbase.gdb'
# Note, no trailing '\\' and this is NOT a workspace, since
# arcpy.env.workspace is not used

out_table = 'my_table'
```

For example, to check for the existence of the table, the following syntax can be used. Note the use of `os.sep`. Make sure to include the `os` module in the `import` line. This is already provided in Chapter 6 demos.

```
if arcpy.Exists(outpath + os.sep + out_table):
    arcpy.Delete_management (outpath + os.sep + out_table)
```

Make sure to pay attention to the parameters that are required to create a table (`CreateTable`) or write features to a feature class (`CopyFeatures`). The feature class does not need to be created when using the `CopyFeatures` routine.

Part I – Create a New Table and Add Fields

Research the ArcToolbox Help (and *A Python Primer*) to create a new table. Make sure to create any necessary variables for the tool. The new table will not use a template that contains the attributes. The code developer will add and define the attribute values. Focus on the required parameters for the ArcGIS tols. Most of the optional parameters are not required.

Make sure to add proper checking for the existence of the new table, so that it can be deleted if it already exists.

1. From within ArcCatalog create a new file geodatabase that will hold the new table. This step will not be done programmatically. Put the new file geodatabase in the **\PythonPrimer\Chapter06\MyData** folder.

2. Create a new table from scratch with the following fields and data types.

    a. LakeFC_ID - short
    b. Lake_Name – text, 50
    c. Lake_Info_ID – long
    d. Lake_Temp – short

3. **Extra** - Create a variable that will hold the file geodatabase feature class name. This will be used to write out the joined records to a feature class. This feature class will need to be written out to the same file geodatabase location as the table.

Part II – Create and Use an Insert Cursor to Write the Joined Attribute Values to an Output Table

1. Create the insert cursor using the following syntax:

```
irows = arcpy.InsertCursor( outpath + os.sep + out_table)
```

2. The insert cursor definition (above) will occur *before the* `for` *loop for the* `srows`.

3. Writing out the new table using insert cursor methods will occur *within the* `for` *loop for the* `srows`.

4. The insert cursor will NOT require a looping structure to write values to the new table.

5. The insert cursor will NOT require a query parameter.

6. The reader will need to use one of the two methods for writing values to the new table:

Method 1

```
row.Output_Field_Name = <variable already defined in
Demo6d.py>
```

For example,

```
row.Output_Field_Name = HydroID
```

Method 2

```
row.setValue('Output_Field_Name', <variable already
defined in Demo6d.py>)
```

For example,

```
row.setValue('Output_Field_Name', HydroID)
```

Notice in the `setValue` there is no equal (=) sign.

7. The respective output fields will be assigned the following values from the joined fields. HINT: Variables for the joined values are already defined in Demo6d.py.

Data Set	Field from Joined Data	Output Field
Lakes (feature class)	Lakes.HYDRO24CA1	LakeFC_ID
Lakes (feature class)	Lakes.HNAME	Lake_Name
Lake_Info (table)	Lake_Info.LakeID	Lake_Info_ID
Lake_Info (table)	Lake_Info.Temp_F	Lake_Temp

8. **Extra** - Write the "joined" data to a feature class. Make sure to use the variable that you defined above. No cursor methods will be involved to write the data to the output feature class. Make sure to add lines to check for the existence of the feature class before writing the data out. The `CopyFeatures` routine will be used in this step.

Chapter 6: Questions

These questions cover the content in Chapter 6, the demonstrations, and the exercise.

1. Briefly describe what each cursor does and an example of when it might be used.

    a. Search
    b. Insert
    c. Update

2. What is the purpose for creating attribute indexes?

3. What feature or table type does a table join require?

4. Describe the parameters required for a table join.

5. Describe how the attribute names change after the table join.

6. Describe how the data can be accessed in a joined table.

From the Exercise

1. What tool is used to create a new table?

2. What tool is used to create new attributes in the table?

3. Name the cursor type used to write out records in the exercise.

4. Describe where in the code the following occur:

    a. Where is the cursor created?
    b. Where are records updated?

Extra

5. If features were written out to a feature class, what tool was used?

6. What do the attributes in the feature class look like? How are the attribute names different from the respective feature class or table attribute names used in the join?

Chapter 7 Describing Data and Operating on Lists

Overview

Part of conducting data analysis with geospatial information sometimes involves knowing some fundamental information about the feature classes, data tables, and images such as the names and data types of attribute fields, the spatial reference system, the different kinds of features in a geodatabase or folder, or the number of bands, rows, columns, and format of an image among others. This kind of information can be used to process certain kinds of data or gain access to certain parts of a feature class or image for additional processing. In addition to accessing these characteristics of a dataset, it is sometimes useful to create a list of values so that processes can be developed to operate over the collection of elements in the list. Lists can contain elements such as fields, feature classes, images, workspaces, and tables among others.

This chapter will focus on obtaining information about the data sets through the use of the Describe function and a variety of list functions to generate a list of specific information and then write coding structures to process specific elements in the list. The reader can refer to the ArcGIS Help documents that refer to **Geoprocessing—Geoprocessing with Python—Accessing geographic data in Python—Describing data** and **Geoprocessing—Geoprocessing with Python—Working with sets of data in Python—Listing data**.

Describing Data

When working with a variety of data formats (feature classes, tables, images, geodatabase, CAD, networks, etc.), especially when writing Python scripts, a code developer may need to access properties that can be used to make decisions for processing. The `Describe` ArcGIS routine provides this ability and will work with all of the formats supported by ArcGIS. For a full account of the data properties refer to the ArcGIS Help in **Geoprocessing—The ArcPy site package—Functions--Describing data.**

To access data properties the `Describe` routine is used which is able to accept many different kinds of properties depending on the data type. Before specific properties can be acquired from the data set, the Describe routine must be used to "describe" or obtain properties about a specific dataset. To do this a variable is assigned to the results of the `Describe` routine of the data set. For example, to access specific properties of an image data set from a satellite sensor, the `Describe` syntax may look like the following:

```
in_image = 'c:\\images\\landsat.img'
desc_img = arcpy.Describe(in_image)
```

Once the `Describe` routine is defined for a given dataset, specific properties of the data set can be accessed and used. In this simple example below the spatial reference, number of sensor bands, and image format are printed to the Python Shell.

```
# Demo 7 - Describing and Listing Data

import arcpy, sys, traceback

arcpy.env.workspace = 'C:\\GIS Programming\\PythonPrimer\\Chapter07\\Data\\'

in_image = 'tm_sacsub.img'

desc_image = arcpy.Describe(in_image)

spat_ref = desc_image.spatialReference.name
num_bands = desc_image.bandCount
img_format = desc_image.format

print 'Spatial Reference: ' + str(spat_ref)
print 'Number of Bands: ' + str(num_bands)
print 'Image format: ' + img_format
```

Since an image is being used as the parameter for the `Describe` routine, properties about images can be retrieved (such as the number of bands or format). See ArcGIS Help in **Geoprocessing—The ArcPy site package—**

Functions--Describing data—Describe properties—Raster dataset properties.
In addition, more general properties can be obtained, such as the spatial reference. See ArcGIS Help in **Geoprocessing—The ArcPy site package—Functions--Describing data—Describe properties—Dataset properties**. Dataset properties are those properties that apply to almost all geospatial datasets such as spatial reference, data type, and spatial extent, among others. Code developers will need to become familiar with the type of data they are using in their scripts and then consult the `Describe` properties that pertain to their needs.

Using `Describe` properties can be used in code to make decisions (such as with the use of the `if` conditional statement). For example, a conditional statement can be used to check to see if a spatial reference is associated with the data set. If one does not exist, then a `print` statement can be used to report this back to the user (or possibly printed to a "log file"). Log files will be discussed in Chapter 8). If a spatial reference does exist, then process can be developed to perform one or more tasks on the data set. The example below shows a conditional statement that checks for a spatial reference. If a spatial reference exists, then another conditional statement is written to see if the image has six bands. If it does, then a statement is written to process an algorithm on the image (the Normalized Difference Vegetation Index (NDVI), in this case). The reader should also note that to perform the NDVI, the Spatial Analyst extension is used. See **Demo7a.py** for the actual script.

```
in_image = 'tm_sacsub.img'
NDVI_image = 'NDVI.img'

desc_image = arcpy.Describe(in_image)

try:

    spat_ref = desc_image.spatialReference.name
    num_bands = desc_image.bandCount
    img_format = desc_image.format

    print 'Spatial Reference: ' + str(spat_ref)
    print 'Number of Bands: ' + str(num_bands)
    print 'Image format: ' + img_format

    if spat_ref <> 'Unknown':

        if num_bands == 6:

            # Process Normalized Difference Vegetation Index (NDVI)

            # (Raster Band 4 - Raster Band 3) / (Raster Band 4 + Raster Band 3)

            NDVI = Float(Raster(in_image + '\\Layer_4') - Raster(in_image + '\\Layer_3')) / Float(Raster(in_image

            print 'Saving ' + NDVI_image

            NDVI.save(NDVI_image)

        else:

            print 'This image does not have 6 bands, \n' + 'the algorithm will not be processed'

    else:

        print in_image + ' does not have a spatial reference'

except:
```

NDVI is an algorithm that has been developed to quantify how much healthy biomass exists in a remotely sensed image.

The original image (**tm_sacsub.img**) and the resulting image from the script (**NDVI.img**) are shown below.

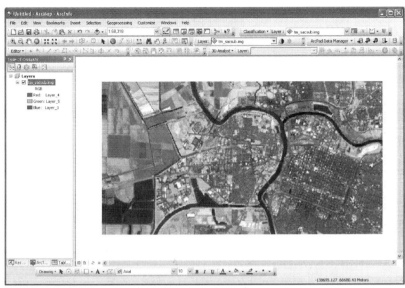

Original Landsat Image subset (tm_sacsub.img).

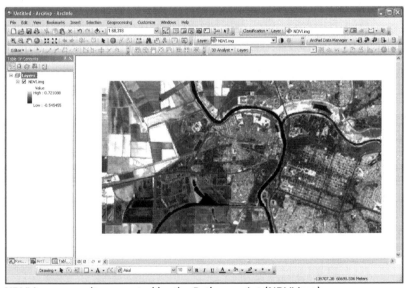

NDVI image result generated by the Python script (NDVI.img).

Once the algorithm is processed, the resulting image is then written to an output file (**NDVI.img** using the `NDVI_image` variable).

If the image data set does not have a spatial reference a print statement is provided to let the user know that one does not exist. In addition, if the image does not have six bands, another print statement is used to let the user know that the NDVI algorithm will not process the image.

Listing Data

Lists are used as primary structures to perform iterative tasks and are at the heart of being able execute "batch processes" (i.e. performing numerous geoprocessing tasks in sequence). To process a list, one must be created. A number of listing routines exist in ArcGIS and can be used in Python scripting. Typically, a list involves developing a specific list of data types. Once a list is created, then individual elements of the list can be processed. For example, a list of shapefile feature classes can be created and then iteratively processed to convert them into file geodatabase feature classes. The ArcGIS Help topic **Geoprocessing—Geoprocessing with Python—Working with sets of data in Python—Listing data** provides an overview of all of the list routines ArcGIS supports as well as links to **Geoprocessing—The ArcPy site package—Functions—Listing data** for help on the specific list functions.

The example below illustrates the use of the `ListFeatureClasses` routine to convert shapefiles to file geodatabase feature classes. See **Demo7b.py** for the actual script.

The general syntax for the `ListFeatureClasses` routine is:

```
arcpy.ListFeatureClasses({wild_card}, {feature_type}, {feature_dataset})
```

1. *wild_card* – indicates a set of characters to help limit the list (e.g. all feature classes beginning with the letter "F")
2. *feature_type* – indicates to limit the list to the type of feature (point, line, polygon, etc). See the ArcGIS Help documentation for a full list
3. *feature_dataset* – indicates that the list can be limited to feature classes found in a feature dataset. Default is empty and will return only standalone feature classes (i.e. those not found in a feature dataset)

Notice that all of the parameters for the `ListFeatureClasses` routine are optional. In many cases, the code developer may choose to create a list of all feature classes in a given workspace. In this case, the following syntax becomes:

`arcpy.ListFeatureClasses()`

This option is shown below in the script.

```
Demo7b.py - C:/GIS Programming/PythonPrimer/Chapter07/Demo7b.py
File Edit Format Run Options Windows Help
import arcpy, os, sys, traceback

arcpy.env.workspace = 'C:\\GIS Programming\\PythonPrimer\\Chapter07\\Data\\'

fgdb_path = 'C:\\GIS Programming\\PythonPrimer\\Chapter07\\MyData\\'

fgdb = fgdb_path + 'Ch07_fgdb.gdb'

try:

    # Create a list of feature classes (from the workspace, in this case a folder)
    fc_list = arcpy.ListFeatureClasses()

    # Process each feature class in the list using a for loop
    for fc in fc_list:       # fc = the shapefile feature class name in the list defined as (fc_list)

        # if the feature class exists in the file geodatabase, delete it

        # fc.rstrip('.shp') strips off the .shp file extension
        # os.sep is used to create the proper full path structure (syntax)
        # to the file geodatabase feature class
        # (e.g. 'C:\\GIS Programming\\PythonPrimer\\Chapter07\\MyData\\Ch07_fgdb.gdb\\<feature_class_name>')

        print 'Checking to see if ' + fc.rstrip('.shp') + ' exists in ' + fgdb
        if arcpy.Exists(fgdb + os.sep + fc.rstrip('.shp')):

            print 'Deleting ' + fc.rstrip('.shp') + ' from ' + fgdb
            arcpy.Delete_management(fgdb + os.sep + fc.rstrip('.shp'))

        # convert the shapefile feature class to a file geodatabase feature class by using
        # the CopyFeatures tool
        # NOTE: The same data path structure as above using the os.sep and rstrip() Python structures

        print 'Copying ' + fc + ' to ' + fgdb
        arcpy.CopyFeatures_management(fc, fgdb + os.sep + fc.rstrip('.shp'))

except:
```

The above script creates a list of feature classes and then uses a `for` loop to iterate over the list, in this case to create a file geodatabase feature classes from the shapefile feature classes using the `CopyFeatures` tool.

Note the structure of the `for` loop. A variable (`fc`) is created to accept an element from the list (`fc_list`). `fc_list` is a variable name that represents the list of feature classes from the `ListFeatureClasses` routine. This variable can then be used in subsequent code. Because the list contains all of the feature classes in the workspace (a folder containing the shapefile feature classes), the code does not need to know a specific name of a feature class and hence can use a variable to reference a specific feature class. Also, as shown in the print and processing statements, the `fc` variable is used in a number of different ways.

Some of new Python scripting syntax examples are used in this script: `os.sep` and `rstrip()`. The `os.sep` routine creates a folder or directory separator character (for Windows OS, it is the "\"). Since the workspace is not defined for the file geodatabase, the following conditional statement uses the `os.sep` routine when checking to see if the feature class exists in the file geodatabase. The `os.sep` Python routine is used to place a "\" between the file geodatabase path and the feature class.

`'c:\GIS Programming\PythonPrimer\Chapter07\MyData\Ch07_fgdb.gdb'`

To append the feature class name (i.e. one of the feature class names from the `ListFeatureClasses` routine) to the file geodatabase path, the `os.sep` routine is used.

`'c:\GIS Programming\PythonPrimer\Chapter07\MyData\Ch07_fgdb.gdb\Sacramento_streets'`

NOTE: The `os.sep` adds the ("\") between **Ch07_fgdb.gdb** and **Sacramento_streets**. In addition, the above syntax actually occurs on a single line. Two lines are used so that it is readable.

In addition, the `rstrip()` routine is used to "strip off from the right side" (the '.shp' extension) from the shapefile name, leaving the root name. Since the script creates a list of shapefile feature classes and the `for` loop uses the fc variable to access a specific element in the list, the syntax for the `rstrip()` routine is:

`fc.rstrip('.shp')`

This syntax takes the value of fc (in this case a feature class name from the list of feature classes) and strips from the right side of the feature class name the string '.shp' (in this case the extension of the shapefile name), which leaves the "root name" of the feature class.

If `fc = 'Sacramento_streets.shp'`, then

`fc.strip('.shp')` yields, `'Sacramento_streets'`

Using both `os.sep` and `rstrip()`, the full syntax for the parameter used in the `arcpy.Exists` function is:

`arcpy.Exists(fgdb + os.sep + fc.rstrip('.shp'))`

which would yield the following if it was completely "hard coded".

```
if arcpy.Exists('c:\\gis
programming\\PythonPrimer\\Chapter07\\MyData\\Ch07_fgdb.fgd
\\Sacramento_streets'):
```

NOTE: The above syntax would typically be shown in a single line in a Python script.

Refer to **Demo7b.py** to review the actual syntax. The ArcGIS Help documentation shows a variation of this method (using the `os.path.join` and the `.strip` Python routines) for the `CopyFeatures` routine to perform featureclass format conversion. The `os.sep` routine is illustrated since it may be required for a variety of geoprocessing routines. Reviewing the `CopyFeatures` routine also represents that other Python syntax is available for use in one's own scripts. The reader is encouraged to review and learn other Python syntax when it is encountered in the ArcGIS Help or from other sources. This will help build the code developer's script repository and techniques.

With a relatively small script that uses a list routine, converting feature classes from one format to another becomes more automated. The script's generic form indicates that this it can be implemented many different times with only a few small changes in the code (primarily, the workspace location of the input feature classes that make up the list and the output file geodatabase location). In addition, the number of feature classes can vary within the folder, so the script can operate on one or even hundreds or thousands of feature classes with the same script.

Summary

Up to this point, the examples have shown only a few actual Python coding structures (e.g. variable names, conditional statements, and looping structures). The code developer may find it useful to implement other Python structures like those shown in this chapter. This will require the developer to research and understand other Python structures and test them in the respective code. ArcGIS Help may provide some clues or hints to these other structures. In addition, the ArcGIS forums are other places to research code and ask questions to the global community of ArcGIS users and developers. These kinds of activities can take significant time when developing Python scripts for geoprocessing. The code developer should keep in mind that one of the primary reasons for developing a script is to automate routine and often manual tasks and so the research and development on a script may end up saving significant time in the long run where the analyst can focus on more important and complex tasks. Also, once coding elements are found and successfully developed, they can become part of the code developer's "code repository" or "portfolio" of script structures and processes that can be used in other programming routines.

As illustrated in this chapter describing data provides some additional specific access to data properties that can be used to make decisions as well as using lists to iterate through groups of data so that common repetitive tasks can become more automated through the use of scripting. The `Describe` and `List` functions allow scripts to become more dynamic so that processing can be implanted on different data types and for different needs.

Demos Chapter 7

Demo7a and **Demo7b** illustrate the tasks described above using the `Describe` and `ListFeatureClass` routines.

The concepts illustrated in the demos and exercises are:

ArcGIS Concepts

Describe
ListFeatureClasses
ListRasters
CopyFeatures
Exists and Delete
Spatial Analyst module
Raster processing
Buffer
Building queries using variables as values
SelectByAttribute
ExtractByMask
Save Rasters
Build Image Pyramids

Python Concepts

`for` loop
`if...else` conditional statements
Indentation
Casting numbers to strings
`os` module
`os.sep`
`rstrip()`
`strip()`
`os.path.join()`
`os.path.splitext()`

Demo 7a: Describe Properties of an Image

The following demonstration shows how to create and use the `Describe` routine for an image and then set some variables for some of the properties of the image. The demonstration code checks to see if a spatial reference exists for the image; if it does, then a second check is performed to see if the image has six bands. If the image has six bands, then the Normalized Difference Vegetation Index (NDVI) algorithm is performed. The Spatial Analyst extension is used, since the algorithm uses band math to perform the raster (image) function. The data can be found in the **\PythonPrimer\Chapter07\Data** folder. An example of the NDVI output image is provided (**NDVI_out.img**). The images and their properties can be viewed in ArcMap.

This demo uses the **tm_sacsub.img** ERDAS Imagine formatted image. This image is a subset of a Landsat TM (Thematic Mapper) satellite image for a portion of Sacramento, CA. NOTE: The workspace may need to be changed for this demo to perform correctly on the user's system.

1. Import the modules required for the program. In addition import the Spatial Analyst module, since this script uses band math, which requires the Spatial Analyst extension. For this demo it is not important to know why this syntax is used with the Spatial Analyst (`sa`) Python module.

    ```
    import arcpy, sys, traceback
    from arcpy.sa import *
    ```

2. Check out the Spatial Analyst extension. Checking out the extension is the method to use so that the functionality of the extension can be accessed within a script. The `CheckOutExtension` routine performs the same function as "checking" the box under the **Customize— Extensions** ArcGIS menu option.

    ```
    arcpy.CheckOutExtension("spatial")
    ```

3. Set up a workspace to a folder that contains the Landsat image

4. Set up some variables, one for the input (**tm_sacsub.img**) and another for the NDVI image (**NDVI.img**) result.

5. Set up the `Describe` routine for the Landsat image. Review the `Describe` routine in the ArcGIS Help for more details describing raster data.

The code should look like the following:

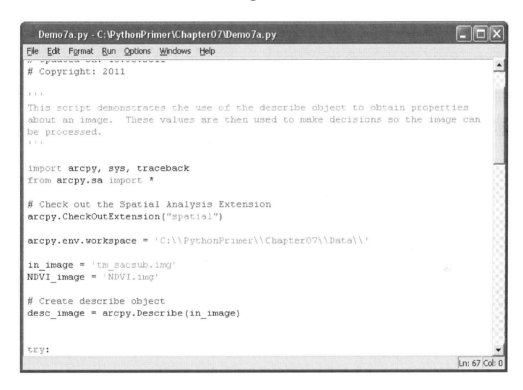

Next create some variables for different properties of the image.

6. Create variables for some of the image properties. This demonstration creates variables for the following. See the ArcGIS Help for more information on the specific properties available for raster (image) data.

    a. Spatial reference
    b. Number of bands
    c. Image format

7. Some print statements are provided to show that values are actually obtained from the different image properties. These are just to illustrate some of the `Describe` values.

8. Create an `if` statement that checks to see if a spatial reference value exists.

9. Create a second conditional statement that checks to see if the number of bands equals 6. Six is the number of functional bands often found with Landsat TM data.

The code should look like the following up to this point.

```
arcpy.env.workspace = 'C:\\PythonPrimer\\Chapter07\\Data\\'

in_image = 'tm_sacsub.img'
NDVI_image = 'NDVI.img'

# Create describe object
desc_image = arcpy.Describe(in_image)


try:

    spat_ref = desc_image.spatialReference.name
    num_bands = desc_image.bandCount
    img_format = desc_image.format

    print 'Spatial Reference: ' + str(spat_ref)
    print 'Number of Bands: ' + str(num_bands)
    print 'Image format: ' + img_format

    if spat_ref <> 'Unknown':

        if num_bands == 6:

            # Process Normalized Difference Vegetation Index (NDVI)

            # (Raster Band 4 - Raster Band 3) / (Raster Band 4 + Raster Band 3)
```

10. Add the specific NDVI algorithm code. NDVI is a common algorithm used with multi-spectral satellite imagery that measures the quantity of healthy biomass. The algorithm can be found in most remote sensing and digital image processing texts. The Python code is provided here for

convenience. Refer to **Demo7a.py** for the exact syntax. NIR refers to the near infrared band collected by the Landsat sensor.

(Landsat NIR Band 4 – Landsat RED Band 3) / (Landsat NIR Band 4 + Landsat RED Band 3)

```
NDVI = Float(Raster(in_image + '\\Layer_4') - Raster(in_image +
'\\Layer_3')) / Float(Raster(in_image + '\\Layer_4') +
Raster(in_image + '\\Layer_3'))
```

NOTE: The above syntax is actually written on a single line in the Python editor.

11. Add a Save line to save the NDVI result to a new image (using the variable defined above).

12. Two `else` statements are written to let the user know that 1) the image does not have six bands and will not be processed and 2) the image may not have as spatial reference associated with it.

13. The `except` block is added. This is the same code that has been used in many of the scripts.

```
num_bands = desc_image.bandCount
img_format = desc_image.format

print 'Spatial Reference: ' + str(spat_ref)
print 'Number of Bands: ' + str(num_bands)
print 'Image format: ' + img_format

if spat_ref <> 'Unknown':

    if num_bands == 6:

        # Process Normalized Difference Vegetation Index (NDVI)

        # (Raster Band 4 - Raster Band 3) / (Raster Band 4 + Raster Band 3)

        NDVI = Float(Raster(in_image + '\\Layer_4') - Raster(in_image + '\\Layer_3')) / Float

        print 'Saving ' + NDVI_image

        if arcpy.Exists(NDVI_image):
            arcpy.Delete_management(NDVI_image)

        NDVI.save(NDVI_image)

    else:

        print 'This image does not have 6 bands, \n' + 'the algorithm will not be processed'

else:
```

Demo 7b: Listing Data

This script creates and uses one of the listing routines found in ArcGIS. In particular, a list of feature classes is created. Any shapefile that exists in the list is converted to a file geodatabase feature class. The Python constructs `os.sep` and `rstrip` are used. See **Demo7b.py** commentary for more details on each of these constructs.

1. Import the proper modules. The `os` module is also included since the `os.sep` routine is used in this script.

2. Set up a workspace for the input path and a variable for the file geodatabase. Set up a variable for the pre-existing geodatabase.

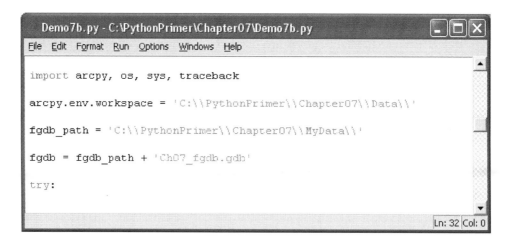

3. Create the list of feature classes using the `ListFeatureClasses` routine.

4. Use a `for` loop to process each of the feature classes in the list.

5. Create an `if` statement to check to see if the feature class exists in the file geodatabase. The `if` statement uses an element (`fc`) from the feature class list (`fc_list`). The `os.sep` character ("\\" for a

Windows OS) is also used in the `if` statement to properly build the path to the shapefile.

```
if arcpy.Exists(fgdb + os.sep + fc.rstrip('.shp')):
```

Note above `fgdb` points to the file geodatabase location (`fgdb_path + Ch07_fgdb.gdb`) which is not a workspace (since `arcpy.env.workspace` is not used). To check to see if the specific feature class exists, the `os.sep` character must be used to form the entire path to the file geodatabase feature class. In addition, the specific feature class can be checked by using syntax from a specific element in the list of feature classes. In this case, since the list contains shapefiles, the specific elements will be in the form *featureclass1.shp*, *featureclass2.shp*, etc. File geodatabase feature classes do not have a file extension, so the *".shp"* needs to be stripped from the specific value in the list. This can be accomplished by using the syntax `<list_element>.rstrip('<characters to remove>')`. Using the syntax in the demo the following is performed.

```
fc.rstrip('.shp')
```

takes an element (`fc`) from the list (`fc_list`) and strips off (`rstrip` – right strip) the *".shp"* extension. The result is just the "root name" of the feature class. For example, *featureclass1.shp* becomes *featureclass1*.

The `Exists` statement would look like this if hard coded:

```
C:\\PythonPrimer\\Chapter07\\MyData\\Ch07_fgdb.gdb + '\\' + featureclass1
```

NOTE: The above syntax actually is written on a single line.

Hence the `Exists` routine can check to see if the feature class exists in the file geodatabase. If it does, then it is deleted which uses similar syntax to the `Exists` routine.

```
fgdb = fgdb_path + 'Ch07_fgdb.gdb'

try:

    # Create a list of feature classes (from the workspace, in this case a folder)
    fc_list = arcpy.ListFeatureClasses()

    # Process each feature class in the ilst using a for loop
    for fc in fc_list:      # fc = the shapefile feature class name in the list defi
                            # fc_list is a variable the contains the full list of fe
                            # from the ListFeatureClasses routine above

        # if the feature class exists in the file geodatabase, delete it

        # fc.rstrip('.shp') strips off the .shp file extension
        # this method can be used to show just the filename
        # the user can also use this syntax with the same effect:  filename, ext =
        # where the filename variable will be the "root name" and ext is a variable
        # ".shp" (or the extension.  An extra line of code would need to be written
        # os.sep is used to create the proper full path structure (syntax)
        # to the file geodatabase feature class
        # (e.g. 'C:\\GIS Programming\\PythonPrimer\\Chapter07\\MyData\\Ch07_fgdb.gd

        print 'Checking to see if ' + fc.rstrip('.shp') + ' exists in ' + fgdb
        if arcpy.Exists(fgdb + os.sep + fc.rstrip('.shp')):

            print 'Deleting ' + fc.rstrip('.shp') + ' from ' + fgdb
            arcpy.Delete_management(fgdb + os.sep + fc.rstrip('.shp'))
```

The script completes by using the `CopyFeatures` to copy the features from the shapefile to the file geodatabase feature class. Notice similar syntax is used in the `CopyFeatures` routine for the output feature class name.

```
            # fc.rstrip('.shp') strips off the .shp file extension
            # this method can be used to show just the filename
            # the user can also use this syntax with the same effect:  filename, ext =
            # where the filename variable will be the "root name" and ext is a variable
            # ".shp" (or the extension.  An extra line of code would need to be written
            # os.sep is used to create the proper full path structure (syntax)
            # to the file geodatabase feature class
            # (e.g. 'C:\\GIS Programming\\PythonPrimer\\Chapter07\\MyData\\Ch07_fgdb.gd

            print 'Checking to see if ' + fc.rstrip('.shp') + ' exists in ' + fgdb
            if arcpy.Exists(fgdb + os.sep + fc.rstrip('.shp')):

                print 'Deleting ' + fc.rstrip('.shp') + ' from ' + fgdb
                arcpy.Delete_management(fgdb + os.sep + fc.rstrip('.shp'))

            # convert the shapefile feature class to a file geodatabase feature class b
            # the CopyFeatures tool
            # NOTE: The same data path structure as above using the os.sep and rstrip()

            print 'Copying ' + fc + ' to ' + fgdb
            arcpy.CopyFeatures_management(fc, fgdb + os.sep + fc.rstrip('.shp'))

except:
```

Exercise 7: Batch Clip Images Using a Feature Class

In this exercise both the `Describe` and `List` routines will be used to "batch clip" a number of images (rasters). The Spatial Analyst extension will also be used. Refer to **Demo 7a** to see how the Spatial Analyst `arcpy` module is imported and how the Spatial Analyst extension is checked out.

NOTES

When working on this exercise, the following should be noted.

1. Make sure to use feature layers where necessary.

2. The output path name is likely needed for the Save routine to save the clipped image. Make sure to include this with the variable that is created for the output image.

3. Use print statements to help troubleshoot and check query syntax. Use ArcMap to help provide clues to query syntax and that the proper polygon is selected.

4. Make sure to use `Exists` and `Delete` properly to help remove pre-existing feature layers and file names.

5. Make sure one step works before working on another when developing this script.

The following files are required for this example and can be found in **\PythonPrimer\Chapter07\Data\images**. Refer to the **Exercise7.mxd** map document.

The images to clip are:

RectifyAA14.tif
RectifyAA18.tif
RectifyBBB19.tif
RectifyQ14.tif

SID_grid.shp – polygon shapefile of grid tiles that will be used in the program

The output to this program will be placed in a folder like:

\PythonPrimer\Chapter07\MyData

Each resulting image will have the format (TIF files):

ClippedAA14.tif
ClippedAA18.tif
ClippedBBB19.tif
ClippedQ14.tif

The following Python syntax will be useful when creating queries, intermediates, and the final output format.

```
tif_root = tif.strip('Rectify.TIF')
```

`strip` will remove any occurrence of `'Rectify'` and `'.TIF'` from the input string. For one of the image files, the result of the above syntax should show a similar result:

AA14
AA18
BBB19
Q14

`tif` (in front of `strip`) refers to a specific TIF image from a list of TIF files.

The script will need to perform the following.

1. Make a list of the rasters that only include the TIFF type. The conditions will be: **'Rectify*'** and be type **'.TIF'**. **NOTE:** In ArcGIS Help it uses the key format word **'TIFF'**. The `ListRasters` routine needs to use the word **'TIF'**.
2. Use a loop to iterate through each element of the raster list to perform the following:

    a. Describe the image and print the following to the Python Shell.
      a. Spatial Reference
      b. Number of Bands
      c. Image format

    b. Create a query variable that uses the "SIDSHT_ID" attribute and uses a variable (e.g. `tif_root`) that will hold the "stripped' value above (e.g. AA14). The `tif_root` variable will be used on the right side of the query. NOTE: Review the values of the SIDSHT_ID field and note that the values in this field contain the "root name" value of the TIF file (but without the "Rectify" or the ".TIF" strings).

    For example, a query will need to be designed such that the syntax has the form:

    "SIDSHT_ID" = 'AA14'

    but without "hard coding" the value 'AA14'. The right side of the equal sign needs to be a variable, yet use the proper syntax. One of the challenges is to make sure the syntax is written correctly for the query.

    c. Use a `SelectLayerByAttribute` routine with the query created above to select a feature (tile or image grid polygon) from the SID_grid.shp file.
    d. Buffer the selected tile (polygon) by 5 feet. Create a buffer file name for each unique buffer, since a buffer will need to be created for each

image in the list. HINT: Use the `tif_root` variable describe above when building the buffer name.
   e. Clip the image with buffered polygon. The step will use the `ExtractByMask` routine under the Spatial Analyst tool set. HINT: Use the the `tif_root` variable described above when building the name for the output image. The image file format will be TIF format.
   f. Refer to the ArcGIS Help for `ExtractByMask` for syntax regarding saving an image. Also refer to **Demo7a.py**.

Extra

Determine what the intermediate file(s) are and clean them up (i.e. delete these files). Make sure to NOT delete the input or output, so be careful.

Build Pyramid Layers for the output images. This was not covered in this chapter, but the reader can research it.

Chapter 7: Questions

1. Reading through the chapter and the ArcGIS Help, what are some of the uses for the `Describe` routine?

2. What are some of the benefits of using the `List` routines? What do these operations allow scripting to accomplish?

3. What do the following do? Refer to *Python.org* or a Python text as needed. Provide an example of each.

    a. `.lstrip`
    b. `.rstrip`
    c. `.strip`

Chapter 8 Custom Error Handling and Creating Log Files

Overview

The reader can see that being able to track and identify errors is very helpful when developing code. So far the demo and exercise programs have relied on a single error block that serves as a default error handler. From a developer or single user's point of view, this may be sufficient; however, if the script will be used by a broader audience, then using more specific error messages may be appropriate.

In addition, the focus of *A Python Primer for ArcGIS* focuses on writing standalone script that automates tasks. Being able to track and log print messages and errors can be helpful, especially if the scripts are complex or when they are automatically executed (as will be illustrated in Chapter 11). Creating log files, custom print statements, and error handlers can assist the programmer as well as the user log, monitor, and track unexpected problems with data or code. Log files and the use of data/time stamp Python routines can assist in the troubleshooting process.

Custom Error Handlers

The reader has already seen and used the `try:` and `except:` blocks to perform some basic error handling. A number of additional error handling methods exist that can expand the error handling capability of a Python script. Some useful methods are:

a. Using of multiple `try: except:` blocks
b. Using nested `try: except:` blocks
c. Using Python functions for different error messages
d. Creating Python classes for different error messages

Of these methods *A Python Primer for ArcGIS* focuses on the last method, since the ArcGIS Help documents and sample code tend to use this method and are readily available to the programmer. For a more extensive discussion on exception or error handling, consult a Python text or the *Python.org* website.

The reader has likely witnessed a variety of error messages and codes, some of which are not very intuitive for troubleshooting and can confuse an end user when an error is encountered. A code developer can add additional error handling code and messages that provide more constructive and informative messages.

To keep existing code logic simple, often a single `try:` and `except:` block are used for general or default error handling purposes. Following this same logic, error "classes" can be created that specifically target potential trouble spots in the code. For example, consider the following possibility.

A code developer performs a `SelectLayerByAttribute` routine that uses a query. One potential pitfall of this routine is the query could have the incorrect syntax or may include values that return no selected features. If subsequent code attempts to use these selected features (such as in the `SelectLayerByLocation` routine), the code will process, but will still end up with no features. Some additional error handling can be added to stop the program and provide a useful message to both the code developer (likely for troubleshooting during code development) and for the end user.

Creating and Using an Error Handling Class

A Python "class" is a computer programming construct for creating and using "objects." An object is a group of code that has a certain functionality that can be called from and used in script at different places. One can think of a class as a function, however, classes differ from functions in that classes can have properties and methods and have a set of hierarchical organization that are not found in functions. A function might be written to perform a calculation operation. The program may want this routine to execute in different places throughout the program. Since this chapter focuses on error handling classes and functions are not thoroughly discussed, the reader is encouraged to consult the Python.org site or text for a more in depth discussion on classes and functions and how they can be created and used for expanding the abilities of a Python script.

To create an error class, the general structure is shown below:

```
class <name of class> (Exception):
    pass
```

The word `class` is a Python keyword to define a Python class structure. The `<name of class>` is any name the code developer chooses (without spaces). `(Exception)` is another keyword that references a Python class to catch exceptions in code and is placed within parentheses. The `pass` keyword (which is indented on the next line of the `class`) tells Python to "do nothing" and acts as a placeholder. The error classes are always defined at the top of a Python script and just after the import modules so the class can be reference at any point in the Python script. In this case, the name of the class will refer to an exception block such as in this example:

```
ErrorHandlingClass.py - C:/PythonPrimer/Chapter08/ErrorHandlingClass.py
File  Edit  Format  Run  Options  Windows  Help

import arcpy, sys, traceback

class NoWorkspace(Exception):
    pass

try:

    arcpy.env.workspace = 'C:\\PythonPrimer\\Chapter08\\xData\\'

    if not arcpy.Exists(arcpy.env.workspace):

        raise NoWorkspace

    print 'Completed Program.'

except NoWorkspace:
    print 'The workspace does not exist.  Check the workspace location.'

except:

    tb = sys.exc_info()[2]
    tbinfo = traceback.format_tb(tb)[0]
    pymsg = "PYTHON ERRORS:\nTraceback Info:\n" + tbinfo + "\nError Info:\n"
    msgs = "ARCPY ERRORS:\n" + arcpy.GetMessages(2) + "\n"

    arcpy.AddError(msgs)
    arcpy.AddError(pymsg)

    print msgs
    print pymsg

    arcpy.AddMessage(arcpy.GetMessages(1))
    print arcpy.GetMessages(1)
```

In this relatively simple example, a workspace is defined. A conditional statement is set up to test to see if the workspace does NOT exist. If the workspace does not exist, then an exception is raised using the `raise` keyword. In this case, a class is used to define a specific class for exceptions. Since the `pass` keyword is used (i.e. no real functionality occurs), the `except NoWorkspace:` block is called and processes the print statement. In the code above, the path stated above does not exist (note `xData` is not an existing folder) and so the code results in an error.

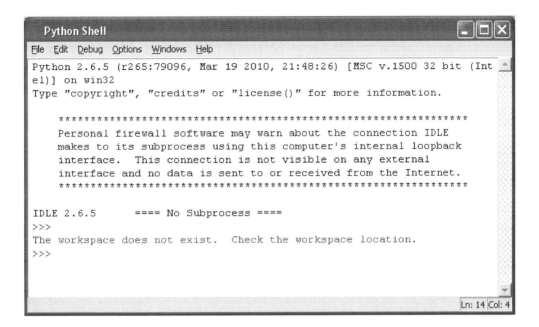

In addition to the above, also note that the except: block is used as well, which serves as a "catch all" for other programming errors.

Writing a custom error handler class helps the code developer to write more specific error messages, the error class can be re-used throughout the code, and additional print statements are not required within the main block of code to provide messages back to the Python prompt when the code fails for the specific checks (e.g. an existing workspace or data path). So, in this example, if the code developer decides to change workspaces, the same kind of check can be performed and call the same error handler.

When developing code, error handlers are often written as a "second phase" of code development. Not until the initial code is written and tested do code developers realize places in the code for error handling. Also the type and quantity of error handlers depends on the end user and audience of the script. A script written and used primarily by the code developer may not warrant specific error handlers. If a script will be deployed so that others can use it, then more specific error handlers will likely be useful.

Using Log Files to Collect Messages

Another useful option for code developers is to print messages to a location other than the Python Shell. If the code developer manually runs a script at a specific time and wants to monitor the messages reporting back from a script, then printing messages to the Python Shell is sufficient. However, most of the methods described in *A Python Primer for ArcGIS* can be automatically run (i.e. without human involvement) and so having a method other than printing messages to the Python Shell is often useful. Printing messages to a file (i.e. a file logging messages and hence the term "log file") is useful from an implementation point of view because the log files can be created during the implementation of the script and can be kept in a repository (i.e. a folder on computer disk) so that they can be audited at any time. Creating a log file is straight forward. Writing messages to the log file only requires a small change in the print syntax.

Creating and Using a Log File

To create a log file for writing messages to it the following is required.

    a. Create a variable for the log file and its path
    b. Open the log file for writing
    c. Close the log file after writing the messages to it

The following script shows the fundamental elements of creating and using the log file.

```
import arcpy, sys, traceback, time, datetime

CURDATE = datetime.date.today()

# The following location (path and folder) must pre-exist on the disk
logpath= 'c:\\PythonPrimer\\Chapter08\\logfiles\\'

# use just the name log.txt for logging messages
logfile1 = logpath + 'log.txt'

# check to see if the log.txt file exists
if arcpy.Exists(logfile1):
    arcpy.Delete_management(logfile1)


log1 = open(logfile1, 'a')

print 'Start logging for log.txt'     # prints message to Python Shell
print >> log1, 'Start logging for log.txt...'  # prints message to log.txt


# Close the log files
log1.close()
```

The above script shows how to create the log file and write to it. Since this book focuses on using Python with ArcGIS the `Exists` and `Delete` functions are used to check whether the log file exists and if it does, deletes it. Alternatively, Python only methods are available to perform file management. The reader can check a Python text or the *Python.org* site for more details.

The first two lines of code set up the path location for the log and then it is used to create the log file. The next line actually opens the file for writing ('a', to append) meaning that the messages created during the current run will be added to a single file.

NOTE: If the `Exists` and `Delete` routine are used to delete the log file before it is re-created, all of the previous messages will be deleted. The code developer will want to consider how log files will be maintained. A naming strategy may need to be developed to keep track of print messages for multiple executions of the script (such as many times per day or log messages to a single file over a number of days). The date and time stamp Python methods mentioned below can assist with developing this strategy.

A variable (e.g. `log1`) is typically used to point to the open file so that print messages can actually be sent to it. Once the file is opened any print statement can be sent to it until the file is closed. To actually write messages to the script notice that the print statement uses '>>' to "redirect" the print statements to the log file using the following syntax. The `log1` variable and the message are separated by a comma (",").

```
print >> log1, 'Start logging for log.txt…'
```

In the above script, the message contains two print statements that are almost identical. The first statement prints to the Python Shell, whereas, the second statement is written to the log file.

Once the script is finished writing messages to the log file, it can be closed by using the following syntax. Note that `log1` is the variable that points to the actual log file.

```
log1.close()
```

Adding the Date or Time to a File Name or Message

An often useful addition to a log file or message is providing the date and time. To monitor scripting progress, the code developer may want to notice the date and time a process starts or completes. To use the `date` and `time` Python functions the `time` and/or `datetime` modules need to be imported. The date and time stamps in print statements can help code developers troubleshoot issues when the scripts are scheduled to automatically run (for example, during off-peak hours). The date can be used as part of the log file name so that unique log files can be generated for a given day.

To access the current date of the computer system, the following syntax can be used.

```
CURDATE = datetime.date.today()   # format YYYY-MM-DD
```

The reader can refer to a Python text or the *Python.org* site for more date formatting options.

To obtain the current date and time and use it as a string (for example as a time stamp in a print statement), the following syntax can be used.

```
time.strftime('%c')    # format is:  MM/DD/YYY HH:MM:SS
```

By slightly modifying the script above, the following syntax takes advantage of the date and time.

```
import arcpy, sys, traceback, time, datetime

CURDATE = datetime.date.today()

# The following location (path and folder) must pre-exist on the disk
logpath= 'c:\\PythonPrimer\\Chapter08\\logfiles\\'

# use the current date in the logfile name
logfile2 = logpath + 'log' + str(CURDATE) + '.txt'

# check to see if the log file with the current date exists
if arcpy.Exists(logfile2):
    arcpy.Delete_management(logfile2)

# Open the log files for writing

log2 = open(logfile2, 'a')

print 'Start logging for ' + 'log' + str(CURDATE) + '.txt'          # prints message to Pyt
print >> log2, time.strftime('%c') + ' Start logging for ' + 'log' + str(CURDATE) + '.txt'  #
                                                        # the log file with
                                                        # the current date in the
                                                        # log file name

# Close the log file
log2.close()
```

Summary

This brief chapter provided some basic Python structures and strategies to work with errors and keep track of print messages so that the programmer and end user of the script have more meaningful ways of troubleshooting code, data, and logic problems. The demo and exercise provide some additional opportunities for the reader to develop these strategies.

Demo 8: Create Custom Error Messages

This demo will add a couple of error handling messages to the script used in **Demo5b** (see the **Demo5b.py** script). A copy of this script can be found in the **\PythonPrimer\Chapter08** folder. Two kinds of errors will be created: 1) check will be to see if any features have been selected for the `SelectLayerByAttribute` routine and 2) check to see if any features are selected from the `SelectLayerByLocation` routine. If either case exists, then an empty feature class will be generated, which is not desirable and will cause the program to stop. This demo will illustrate the use of creating error "classes". A number of different options exist with Python; however, ArcGIS preferentially uses the class method. See comments within the demo code for options to test the error messages. The **Demo8.py** script contains the code from the **Demo5b.py** script used in this illustration.

The concepts illustrated in the chapter and demos are:

ArcGIS Concepts

GetCount
Spatial Analyst extension
CheckOutExtension

Python Concepts

`class`
`except`
Unique error handling
Cast numbers and values to strings
`If` statements
Open, close, write to files
`time` and `datetime` modules and formatting
Redirect print statements to files

1. Start by defining an error class. Two will be defined as shown below.

```
print 'Demo 8 - Writing Special Error Messages\n'

import arcpy, sys, traceback

class NoAttributeFeatures(Exception):
    pass

class NoLocationFeatures(Exception):
    pass

arcpy.env.workspace = "C:\\PythonPrimer\\Chapter08\\Data\\"
```

2. Next, toward the bottom of the code, add the specific exception blocks. In this case there are two, since two classes are defined above.

```
    print "Copied selected attributes from " + feat_class + " to " + out_stree

    print "Completed Script"

except NoAttributeFeatures:
    print 'There are no selected features by attribute.\n'\
          'Review the query syntax.'

except NoLocationFeatures:
    print 'There are no selected features by location.\n'\
          'Review the feature layers and the selection type.'

except:

    tb = sys.exc_info()[2]
    tbinfo = traceback.format_tb(tb)[0]
    pymsg = "PYTHON ERRORS:\nTraceback Info:\n" + tbinfo + "\nError Info:\n
    msgs = "ARCPY ERRORS:\n" + arcpy.GetMessages(2) + "\n"

    arcpy.AddError(msgs)
    arcpy.AddError(pymsg)

    print msgs
    print pymsg
```

3. Within the body of the code, add an `if` statement to check to see if the selected feature count is zero. If it is, then "raise" an error using the name of the specific error message. In this case, since the number of selected features is required, the `GetCount` routine is used and will also be used in the `if` statement. Also note that the result value from the `GetCount` is cast as a string (using `str()`) to work properly with the `if` statement.

A second `if` statement will be used to check to see that some polygon features are selected.

To test the first error message, deliberately change the query string to a street "CLASS" that is not available (e.g. "CLASS" = 'X'). After making this change, execute the script to see that the first error message is written to the Python Shell.

To test the second error message, the reader will note that a `SelectLayerByAttribute` routine with the "CLEAR_SELECTION" selection type is provided so that the error message can be tested. If all of the polygon features are "cleared" from the selection, then the `GetCount` result for `result2` will report the total number of features, thus indicating that "no" polygons are selected.

NOTE: If no features are selected, `GetCount` reports the total number of features of the feature class even though there are no selected features.

A test (refer to the second `if` statement shown after the `result2` variable) can be performed to see if the total number of polygons is the same both before and after the selection routines. If they are the same, this will indicate that "no" polygon features are selected, thus "raising" the `NoLocationFeatures` error. Refer to the **Demo8.py** script to review and test the entire script.

The reader can comment out the line containing:

```
arcpy.SelectLayerByAttribute_management(poly_layer, "CLEAR_SELECTION")
```

This will bypass the `NoLocationFeatures` error and execute the `CopyFeatures` routine for the polygon layer.

Exercise 8: Create Custom Error Messages and Log File for a Script

Using the code developed for Exercise 7, add the following options to the code. NOTE: The input paths for the image data will continue to point to **\PythonPrimer\Chapter07\Data** and the output will continue to point to **\PythonPrimer\Chapter07\MyData**. The **logfiles** folder will be added to the **\MyData** folder. See below. The **Exercise7_Batch_Clip_Solution.py** script has been provided in the **\Chapter08** folder that can be used in this exercise.

1. Add a unique error message handler to monitor an error when the number of selected features is zero.
2. Add a unique error message handler to monitor an error when the image does not have a spatial reference (i.e. spatial reference is `Unknown`). `Unknown` is the string name used to indicate that a spatial reference does not exist with a data set.
3. Create a folder under **MyData** called **logfiles**.
4. Create a unique log file that uses the date in the log file name.
5. Write `print` statements to the log file as well as to the Python Shell. Use the time stamp string method shown in the chapter. For example, if the print statement is already created, add additional statements that will be redirected to the log file (so that one message is written to the Python Shell and a second message is written to the log file). Two sets of print statements will result in the code: One is printed to the Python Shell, the other printed to the log file.

NOTES

1. Make sure to import the `time` and `datetime` modules to the script, otherwise the date and time values cannot be used properly. Also make sure to cast the dates and/or times appropriately. See the chapter for more details.

2. Make sure to add all of the required code pieces for creating error handlers.

Chapter 8: Questions

1. What are reasons why code developers would want to create unique error messages?

2. What are log files useful for?

3. What is the benefit of using date and time stamps in log files and within `print` statements?

Chapter 9   Mapping Module

Creating maps and map books can be a time consuming task for a GIS analyst. Until ArcGIS 10 many analysts created a map document by adding specific map elements such as a title, legend, North arrow, and scale bar among others. As any cartographer can attest, creating a map takes skill in using the software to perform the tasks, but also needs an "artistic eye" to produce a map that is appealing to the audience and communicates the intended message. For some cases, spending significant time creating and designing a map for a single geographic area may be warranted. However, if the same kind of map is required for different geographic locations where only the surrounding map elements change (such as the title, subtitle, legend items, date, map sheet number, etc.), making these minor changes can be very time consuming and often involve the analyst creating many separate map documents or manually changing the map elements for a single map document, and then printing or generating individual PDF documents.

Past versions of ArcGIS provided some tools and sample code to assist with the map production efforts of analysts (such a map production sample scripts and third party tools to semi-auto generate map sheet boundaries and map sheet labeling schemes). Creating useful map production tools required the use of Visual Basic for Applications or Visual Studio to create graphical user interfaces as well as special geographic data for making the map production process worthwhile. Examples of a custom mapping automation tool using VBA can be found on the author's website:

www.jenningsplanet.com

Skill Set – Application Development
Gallery – Carto—Solid Waste Route Books or Apps--VBA

Beginning with ArcGIS 10, a new Python module, the `mapping` module, is available to customize map documents as well as provide the ability to "automate" map production. Using Python scripting reduces the coding overhead (i.e. the number of lines of code) often found with Visual Basic or

Visual Studio. In addition, the `mapping` module also contains methods to generate PDF documents or print to available printers. The introduction of the `mapping` module is one of the recent major improvements to the ArcGIS functionality. Additional improvements with automating map production are the introduction of Data Driven Pages and ArcGIS reference material for building map books. See the ArcGIS Help topics under **Mapping and Visualization—Automating map workflows—Data driven pages** and **Mapping and Visualization—Creating a map book**. The reader may find these resources useful when developing a map production strategy and some of them require some additional interaction with ArcMap and data preparation. The focus of this chapter is to illustrate how to use Python to modify some of the map elements that may be present on a map (or map book). This chapter also focuses on the fundamental methods, properties, and functions found in the mapping module which can be used in more complex Python scripts for producing map books. The same maps that are shown on the author's website can be fully reproduced using Python instead of VBA, which involve a single ArcMap document where the script produces 4 sets of maps (overview, vicinity, street detail, and street detail with parcels). The script strategically turns layers on and off, changes the zoom extent, performs definition queries, and modifies the labels and scale bar. The major differences between the concepts in this chapter and those found in a more complex script is the use of looping structures, the use of a tailored map grid, and making sure the queries and layer visibility is performed at the right place in the code.

Overview

Two "views" of geographic data exist in ArcGIS: 1) Data View and 2) Layout View. In Data View, geographic data is viewed, symbolized, and managed without respect to a specific map page size or layout. Data View is typically used when an analyst needs to conduct geographic analysis, review results, and perform other analytical tasks on the data. In Layout View, the geographic data is presented on a map page with a specific size and layout. The Layout View provides the cartographer a (*What You See is What You Get, WYSIWIG*) view of the data to facilitate presentation. The cartographer or analyst will spend time in the Layout View refining the map elements, symbolizing the map with colors, line types, shading as well as the other map elements such a font size, font type, and placement of the elements (on the side, at the bottom, within the map area itself, etc.).

Map Elements

A typical map in ArcGIS consists of the following basic components.

a. *Map document* – the specific map file
b. *Map frame* – the area that contains geographic data
c. *Layers* – data layers found in the table of contents
d. *Legend* – usually a list of data layers and their symbols shown in the map frame
e. *Title* – the theme of the map
f. *Subtitle* – (sometimes optional, depending on the map)
g. *North arrow* – to indicate which direction is North in the map
h. *Scale bar* – scale to indicate the level of detail and distance measurements
i. *Title block* – an area where a cartographer can place information about the map, such as an author, map date, map sheet numbers, disclaimers, map data sources, notes, or other documentation

The following figure shows a typical map with the elements listed above.

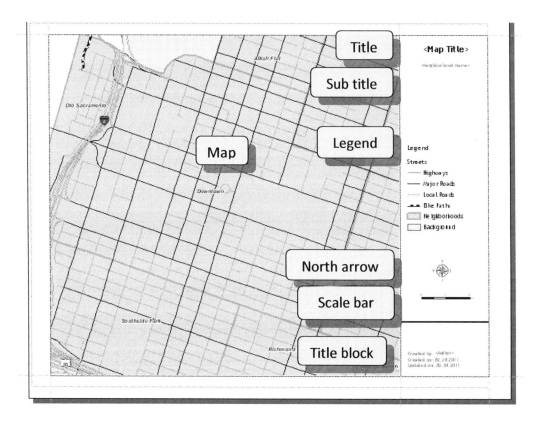

Within ArcMap geographic data is added to the Table of Contents as "layers." The data is also added to the active "data frame." A data frame is "active" when the name is shown as "bold" in the Table of Contents. Although the data frame is distinctly different from the feature classes listed in the Table of Contents, Esri decided to assign the default data frame the name "Layers." Shown below is the table of contents of the map shown above.

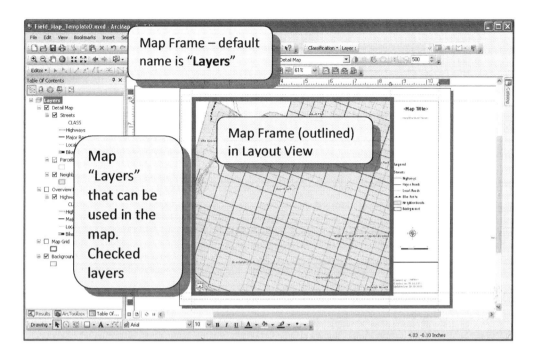

In addition to the data frame and the layers in an ArcMap session, other map elements such as a title, legend, North arrow, scale bar, and other text elements can be added and modified. A cartographer typically adds these elements as required for the type and purpose of the map. The position, font type and size, and other properties of each element can be modified. In ArcGIS, these elements are collectively referred to as "Layout Elements." Depending on the type of layout elements, different kinds of properties are associated with each element. The element types available in ArcGIS are:

a. *Data Frame* – a specific data frame
b. *Graphic* – graphic elements such as a neat line, markers, lines, etc
c. *Legend* – a legend
d. *Mapsurround* – North arrows, scale bars, and scale text (elements related to the map frame)
e. *Picture* – logos, picture graphics added to the map that are not related to geographic data
f. *Text* – any text element, such as a title, subtitle, date, author, map sheet number, and other text

The figure below shows most of the layout elements above and their associated element type.

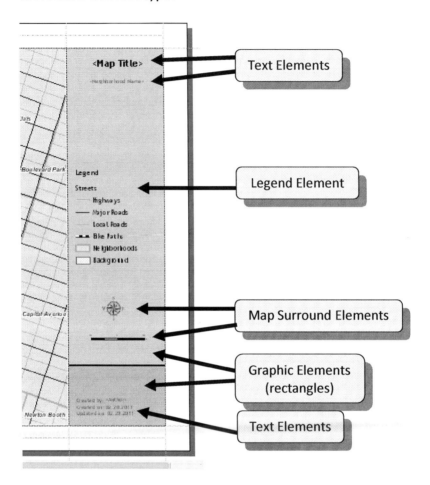

Map Element Relationships

The general relationship among the elements of a map is shown below. The chart indicates some of the fundamental map layout elements and their associated operations. For a full account of the map layout elements available in the mapping module, see the ArcGIS Help documents under **Geoprocessing—The ArcPy site package—Mapping module**.

Map Element

Operations

The primary element is the ArcMap document. Map documents contain data frames, and layout elements and can also be saved, printed, and exported. Layers are typically related to a data frame and can be added, removed, and modified (for example, the symbology can be changed). In addition, data frames can be modified (such as the rotation, scale, units, and dimensions of the map frame on the map layout). The extent of the map frame can be obtained or set explicitly or changed by zooming to selected features. Specific layers can have their appearance changed (such as the visibility, transparency, and labeling) and definition queries can be applied to the map layout. Finally, the layout elements (such as text, graphics, pictures, and legends) can have their position and values set. In addition to the chart above, refer to **Appendix 9.1** for a listing of specific functions, classes, properties, and methods that will be discussed in this chapter.

Prerequisites

Before working with the `mapping` module and manipulating it programmatically, a map document must exist on computer disk and contain at least one map frame. Typically, this map frame will have some data layers and typically include some map elements that are likely to change with each unique map produced. Essentially, this map document can be thought of as a template for an automated map production process. The elements that will change (such as the map title, subtitle, date, legend, North arrow, scale, and various text elements) will also need to be uniquely named through their respective element properties.

Map Design

1. Create a map template (.mxd)
    a. Determine map size
    b. Add data frame(s)
    c. Set up and symbolize Layers
2. Add and name map elements (as required)
    a. Title/sub title
    b. North arrow, scale bar, legend, title block

Create a Map Template

Fundamentally, before working programmatically with a specific map document (and learning the programming fundamentals), it is helpful to create a "template" that contains all (or most of) the data layers required for the map, the page layout (map size and orientation), the map frame, and the organized map layout. This "template" is not the traditional ArcMap template (.mxt), but rather a template (.mxd) that contains a number of common map elements that can be manipulated through scripting methods as part of an automated map production task. Only layers can be programmatically added to the map document, so essentially all of the map elements exist in the map document that will be programmatically modified. The examples in this chapter will illustrate a number of these programming methods using an existing map document with data layer and typical map layout elements. See the ArcGIS Help under **Geoprocessing—The ArcPy site package—Mapping module** to see a full list of the mapping module functionality.

Name Map Layout Elements

In addition to creating a map template, the cartographer or analyst will need to assign unique "names" to each element that is to be manipulated by Python script. This is performed within ArcMap as part of the original map design. Map layout elements can be named by accessing the element's properties (right-click on a specific layout element and choose Properties). The following figure shows an example.

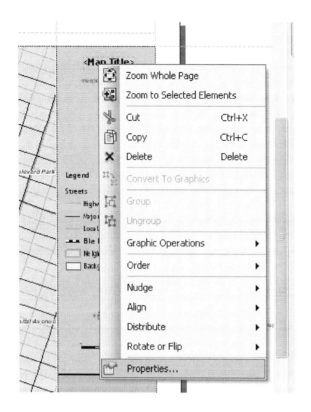

Click on the Size and Position tab to access the layout Elements Name property. In this case, the "name" of the selected layout element is called "Map Title." When the Element Name is changed, the Property dialog box title is also renamed.

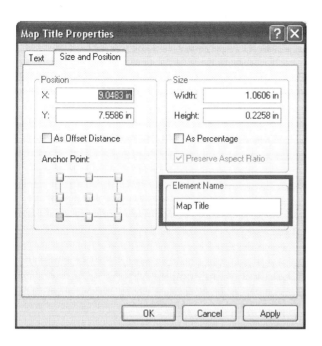

Note the Element Name is different than the actual text shown on the map. The text shown on the map is found under the Text tab in element's properties. Also, depending on the type of layout element, different tabs and properties are available.

The unique name for each layout element is used by the mapping module to gain access to specific elements and programmatically manipulate them. If a unique layout element name does not exist, then it cannot be changed through Python scripting methods.

Also, data layers should have meaningful names in the table of contents. If multiple data frames exist on a single map, they should also be uniquely named. Providing names to layers and data frames can assist with developing Python script to make changes to the map layout. NOTE: Accessing specific data layers and data frames does not require unique names, but it is good practice to use unique names.

Mapping Module Class Properties and Methods

Before discussing specific Python methods to manipulate the map layout the reader should note that many of the coding operations related to the ArcMap document is to access and modify properties and methods from specific mapping classes. A class is a programming term that means a collection of properties, methods, and functions are related to one another. For example a driver may have a car (class) that is a certain make and model, has four wheels, and is white (properties). In addition, the car may have methods that the driver can use to determine the how well the car functions. For example, the drive may have a method called FuelGauge that reports the current level of fuel in the car. It may also have an Odometer method to report the total distance the car has driven or a TripOdometer to report the mileage of the current trip.

From an ArcGIS perspective all of the properties and methods related to each other can be found in their respective classes. Data Frame properties and methods can be found in the Data Frame class. Data frame properties can include the name, scale, rotation among others. These properties can be used to determine or set the data frame name, current scale, or the data frame rotation. Data Frame methods include actions such as zoomToSelectedFeatures that can be implemented to perform an action (in this case, zoom the data frame to the extent of the selected features). Layers and specific layout elements also have their own respective classes that a code developer can use to obtain and manipulate values for each part of a map document. The full list of the mapping module classes (which include descriptions of the properties and methods) can be found under **Geoprocessing—The ArcPy site package—Mapping module—Classes.**

Mapping Module Functions

In addition to properties and methods, the mapping module also has a variety of functions to gain access to different parts of the ArcMap document and its map elements (data frames, layers, layout elements) and for exporting and printing map documents. Mapping module functions are implemented in similar ways to other ArcGIS functions and geoprocessing tools. A mapping module function performs an action to do something with a data frame, layer or layout element. A number of list routines exist that can be used to iterate through layers (for example, to change the visibility in the map) or layout elements (for example, to change the value of text elements). A full list of mapping module functions can be found under **Geoprocessing—The ArcPy site package—Mapping module—Functions.** The ArcGIS function help is divided between **Managing Documents and Layers** (i.e. those related to the ArcMap document contents) and **Exporting and Printing**.

Implementing Map Documents, Data Frames, Layers, and Layout Elements

Now that some of the map document, data layer, and map layout elements fundamentals have been discussed, this section illustrates how to work with these elements programmatically using Python and the mapping module. The examples shown below can be followed in the **Demo 9a Mapping_Module_Overview.py** script and the **Mapping_Module_Overview.mxd**. In addition, the reader should consult the ArcGIS Help documents under **Geoprocessing—The ArcPy site package—Mapping module** which contains all of the documentation related to the mapping module. Refer to specific mapping module classes, properties, methods, and functions as needed.

The general workflow when working with the mapping module is:

1. Import the mapping module
2. Access a specific map document (.mxd)
3. Access a data frame from a list of data frames (often only one)
4. Access and modify layers within the map (as needed)
5. Access and modify layout elements in the map (as needed)
6. Print, export, or save the map

Import the mapping Module

To begin programming Python scripts to work with map documents, the mapping module must be imported.

Notice the 'from arcpy.mapping import *' line. Using this syntax tells Python that the entire functionality of the mapping module will be imported and helps to shorten some of the mapping module related syntax.

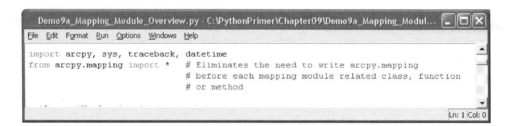

Accessing an ArcMap Document

Next, the code developer needs to access the pre-existing map document. Notice that a variable is assigned the path (folder, not a workspace) for the location of the map document. To access a specific ArcMap document, a path and map document file name is required, not a workspace).

```
Demo9a_Mapping_Module_Overview.py - C:\PythonPrimer\Chapter09\Demo9a_Mapping_Module_...
File  Edit  Format  Run  Options  Windows  Help

import arcpy, sys, traceback, datetime
from arcpy.mapping import *    # Eliminates the need to write arcpy.mapping
                               # before each mapping module related class, function
                               # or method

author = 'N. Jennings'

# reformat date to MM.DD.YYYY
# NOTE: lowercase %y will result in MM.DD.YY format
# See Python.org or text regarding the datetime module

CUR_DATE = datetime.date.today().strftime('%m.%d.%Y')

try:

    # Get the map document.
    # In this case a custom template set up for map production
    # change the paths as needed to run the script form a local system
    datapath = 'C:\\PythonPrimer\\Chapter09\\'
    mappath = datapath + 'MyData\\Maps\\'     # output for PDFs
    mxd = MapDocument(datapath + 'Mapping_Module_Overview.mxd')

    # Report some of the Map Document Properties
```

The specific ArcMap document is assigned to the variable `mxd`.

NOTE: The specific mapping module functionality is accessible by `import arcpy`, however, without `from arcpy.mapping import *`, all of the mapping module specific classes, functions, and methods would need to include `arcpy.mapping` in front of the specific class, function, or method. For the example, without `from arcpy.mapping import *`, the syntax would be written like this for the map document:

`mxd = arcpy.mapping.MapDocument(datapath + 'Field_map_Template.mxd')`

Once the map document is referenced, some map properties can be obtained. Below, see some of the map properties that are printed to the Python Shell.

```
# In this case a custom template set up for map production
# change the paths as needed to run the script form a local system
datapath = 'C:\\PythonPrimer\\Chapter09\\'
mappath = datapath + 'MyData\\Maps\\'    # output for PDFs
mxd = MapDocument(datapath + 'Mapping_Module_Overview.mxd')

# Report some of the Map Document Properties

print 'Map Document Title: ' + str(mxd.title)
print 'Map Document Author: ' + str(mxd.author)
print 'Map Document Summary: ' + str(mxd.summary)
print 'Map Document Desription: ' + str(mxd.description)
print 'Map Document last Date Saved: ' + str(mxd.dateSaved)
print 'Is the Map Document Relative Path checked? : ' + str(mxd.relativePaths)
```

A programmer can use the map properties to make checks on map documents and determine if further changes and modifications are required. For example, a conditional statement can be written to check the last saved date of the map and then make further updates to it if they are needed.

Accessing a Data Frame

Next, a specific data frame can be accessed using the `ListDataFrames` function. The syntax below tells Python to find the data frame named "Layers" from the first data frame in the **Mapping_Module_Overview.mxd**. The `ListDataFrames` function returns a Python list which begins with an index value "0" or first position in the list. A Python list contains a list of specific elements, in this case data frames. Elements in a Python list are accessed by their position in the list, which begins with position (0). The first (and only element) in this example is the only data frame in the map and thus the syntax is:

```
dataframe = ListDataFrames(mxd, "Layers") [0]
```

To access a specific element in a Python list, its index position must be used. In the example below, since the code developer already knows the contents of the specific map document and that it contains only a single map frame, the map frame Python list position (i.e. 0) can be "hard coded" and assigned to the variable dataframe. Like the map document properties shown above, data frame properties can be accessed. Some of them are shown below.

```python
print 'Map Document last Date Saved: ' + str(mxd.dateSaved)
print 'Is the Map Document Relative Path checked? : ' + str(mxd.relativePaths)

# Get a list of data frames
# [0] inicates the first (or only data frame, in this case)
# The data frame name is "Layers", the default name for the
# primary data frame
dataframe = ListDataFrames(mxd, "Layers") [0]

# Report some properties about the map frame named "Layers"

print 'Map Frame Map Units: ' + str(dataframe.mapUnits)
print 'Map Frame Scale: ' + str(dataframe.scale)

# See ArcGIS Help under 'spatialreference' (all one word) for more information
print 'Map Spatial Reference: ' + str(dataframe.spatialReference.name)
print 'Map Frame Anchor Point X Position (page units): ' + str(dataframe.element
print 'Map Frame Anchor Point Y Position (page units): ' + str(dataframe.element
print 'Map Frame Width (page units): ' + str(dataframe.elementWidth)
print 'Map Frame Height (page units): ' + str(dataframe.elementHeight)

# Get the extent of the data frame and assign it to a variable
mapExtent = dataframe.extent

# The following syntax prints out the Min and Max X and Y values
# for the map extent of the data frame.
# Search on Extent in the ArcGIS Help to find more details on
# the extents XMin, YMin, XMax, and YMax

# The string formatting can be found in a Python text or at Python.org
# %f indicates a decimal value will be added to a string
# each %f below is substituted with the appropriate mapExtent.<value>

print 'Map Extent of Data Frame is: \n' + \
      'XMin: %f, YMin: %f, \nXMax: %f, YMax: %f' % \
          (mapExtent.XMin, mapExtent.YMin, mapExtent.XMax, mapExtent.YMax)
```

Accessing Layers

To access the layers in a map, the following syntax can be used:

```python
print 'Map Extent of Data Frame is: \n' + \
    'XMin: %f, YMin: %f, \nXMax: %f, YMax: %f' % \
    (mapExtent.XMin, mapExtent.YMin, mapExtent.XMax, mapExtent.YMax)

# Get a list of layers in the table of contents of the map document
TOCLayers = ListLayers(mxd)

print 'Processing layout elements...'

# loop through the layers
for TOCLayer in TOCLayers:
    print 'Layer Name: ' + str(TOCLayer.name)

    print 'Longname: ' + str(TOCLayer.longName)    # longName provides
                                                    # the ability to use
                                                    # "Group Layers"

    # Detail Map is the group layer
    # Neighborhoods is a layer within the group

    if TOCLayer.longName == 'Detail Map\Neighborhoods':

        TOCLayer.transparency = 50   # 50% transparency

    if TOCLayer.longName == 'Detail Map\Streets':

        TOCLayer.showLabels = True   # turn labels on

    if TOCLayer.longName == 'Detail Map\Parcels':

        TOCLayer.visibility = False  # turn layer off
```

The `ListLayers` function is used to access a list of data layers in the ArcMap document. The resulting list is a Python list and operates similarly to the `ListDataFrames` function. In this case the Python position [0] is not used because a `for` loop is implemented to iterate through all of the layers in the list (that occurswithin the map document). Note that this `ListLayers` function is not referencing a specific data frame, so the resulting list will include all of the layers in the table of contents. The `ListLayers` routine does have an optional data frame parameter that can be used to provide a list of layers associated with a specific data frame. The `for` loop iterates through the list to perform additional operations on different layers. The example above shows how a list of layers is created from a map document (assuming that only one data frame exists in the map document).

```
TOCLayers = ListLayers(mxd)
```

The above script could also have been written by also referencing the data frame parameter.

```
TOCLayers = ListLayers(mxd, '', dataframe)
```

    `mxd` – variable for the map document
    `''` – optional wild card parameter (not used in this example)
    `dataframe` – variable for the data frame

The `for` loop above shows an example of how a layer property can be obtained, in this case the layer's name. Since a loop is being used to iterate over all of the layers in the list, a specific layer's name (or other property) can be accessed by using the variable pointing to a layer in the list and the property. The general form of accessing a layer's property is:

```
Layer.property
```

See the Layer class in the ArcGIS Help under **Geoprocessing—The ArcPy site package—Mapping module—Functions** for a full list of layer properties.

The example above also shows how a layer can be accessed within a group layer. A group layer in ArcMap often contains related layers that may relate to a specific theme or purpose. For example, a "Roads" group layer can contain different road layers such as Highways, Surface Street, and Trails. Similarly, a "Detail Map" group layer, may contain Streets, City Boundary, and Parcel layers. To access a group layer the `longName` property is used. If a layer is in a group layer, the general syntax is:

```
Layer.longName = '<GroupLayer>\<LayerName>'
```

where the `GroupLayer` and `LayerName` are separated by a "\". In the figure above, the Neighborhoods layer is accessed this way:

```
TOCLayer.longName = 'Detail Map\Neighborhoods'
```

Using Layer Properties in the Script

Typically, once layer properties are retrieved, the code developer wants to do something with the values (besides just printing them out to the Python Shell). The use of the property depends on the property values, when the value is obtained and how the coding structure is designed. In the figure below a conditional statement is used to check to see if the `longName` is the "Detail Map\Neighborhoods" layer. If it is, then the layer's `transparency` property is set to 50% (i.e. `TOCLayer.transparency` = 50). Likewise, other properties can be manipulated.

```
print 'Map Extent of Data Frame is: \n' + \
      'XMin: %f, YMin: %f, \nXMax: %f, YMax: %f' % \
      (mapExtent.XMin, mapExtent.YMin, mapExtent.XMax, mapExtent.YMax)


# Get a list of layers in the table of contents of the map document
TOCLayers = ListLayers(mxd)

print 'Processing layout elements...'

# loop through the layers
for TOCLayer in TOCLayers:
    print 'Layer Name: ' + str(TOCLayer.name)

    print 'Longname: ' + str(TOCLayer.longName)   # longName provides
                                                  # the ability to use
                                                  # "Group Layers"

    # Detail Map is the group layer
    # Neighborhoods is a layer within the group

    if TOCLayer.longName == 'Detail Map\Neighborhoods':

        TOCLayer.transparency = 50   # 50% transparency

    if TOCLayer.longName == 'Detail Map\Streets':

        TOCLayer.showLabels = True   # turn labels on

    if TOCLayer.longName == 'Detail Map\Parcels':

        TOCLayer.visibility = False   # turn layer off
```

The reader should recognize the progression or hierarchy of accessing layer properties. To access a specific layer's properties, the map (and optionally the data frame) must be known so that a specific layer can be obtained; only then can layer properties be accessed and used for various purposes. As show above, the script accesses a specific ArcMap document (`Mapping_Module_Overview.mxd`). Since the map document only contains one data frame, the code developer only needs to use the map document as the parameter in the `ListLayers` routine (i.e. `ListLayers(mxd)`). Once the list is obtained, specific layer properties can be accessed and modified as needed. The same will be true when the code developer wants to manipulate the layout elements of a map, which will be shown below.

Accessing and Changing Layout Elements

To manipulate the layout elements a similar process is used. The layout elements are accessed through the use of the `ListLayoutElements` function. Since several different kinds of layout elements exist, one of the specific layout elements types is often used in the `ListLayoutElements` function. The specific element types are:

TEXT_ELEMENT – text elements in the map
LEGEND_ELEMENT – a legend, if used in the map
DATAFRAME_ELEMENT – any data frame in the map
GRAPHIC_ELEMENT – graphic elements (boxes, circles, and other shapes)
MAPSURROUND_ELEMENT – north arrows, scale bars
PICTURE_ELEMENT – image files (e.g. logos)

See **Appendix 9.1** or the ArcGIS Help documentation regarding layout elements for more information.

The script below shows how the text elements (element type is `'TEXT_ELEMENT'`) of a map can be accessed and used.

```
if TOCLayer.longName == 'Detail Map\Parcels':

    TOCLayer.visibility = False   # turn layer off

# Get a list of "text" type layout elements from the map document
tElements = ListLayoutElements(mxd, "TEXT_ELEMENT")

print 'Processing text elements...'
for tElement in tElements:

    # if the text element name is 'Map Title',
    # then assign a specific name for the title
    # see the elements properties in ArcMap (under size and position)

    if tElement.name == 'Map Title':

        tElement.text = 'Test Map'
        tElement.elementPositionX = 8.7  # Anchor point is lower left
                                         # The anchor point is used
                                         # to locate the lower left
                                         # page position of the text

        # The X and Y positions (and Height and Width
        # can be found in the element properties
        # of the layout object under Size and Position

    if tElement.name == 'Print Date':

        tElement.text = str(CUR_DATE)  # assign the current date
                                       # to this text element
```

Shown above, the `ListLayoutElements` function is used to obtain only the "text" type elements from the map document (`ListLayoutElements(mxd, 'TEXT_ELEMENTS'`). All of these happen to be part of the title block in the ArcMap document (see the **Mapping_Module_Overview.mxd** associated with this chapter). A `for` loop can be used to iterate through many of the text elements and make the required changes using the specific text element properties. Remember that the layout elements use the "Element Name" described above so that Python and the `mapping` module can access specific layout elements. Only those elements with unique element names can be modified programmatically. Elements that are not expected to be modified programmatically do not need to have a unique element name.

The above script shows that the text element with the element name "Map Title" will have its "Text Property" assigned to the string "Test Map." The text will also be placed at the page layout's X position at 8.7 inches from the left edge of the page. Also, the text element with element name "Print Date" will have its text property assigned to the current date of the computer system. The script has a variable called CUR_DATE that is assigned the current computer system's data. This variable is defined towards the top of the script. Note in the **Mapping_Module_Overview.py** script that some special formatting is provided so that the date will have the format MM.DD.YYYY.

```
Demo9a_Mapping_Module_Overview.py - C:\PythonPrimer\Chapter09\Demo9a_Mapping_Mod...
File  Edit  Format  Run  Options  Windows  Help

import arcpy, sys, traceback, datetime
from arcpy.mapping import *   # Eliminates the need to write arcpy.mapping
                              # before each mapping module related class, function
                              # or method

author = 'N. Jennings'

# reformat date to MM.DD.YYYY
# NOTE: lowercase %y will result in MM.DD.YY format
# See Python.org or text regarding the datetime module

CUR_DATE = datetime.date.today().strftime('%m.%d.%Y')

try:
```

Exporting Map to PDF or Print Map to Printer

A final step to modifying the map is to export or print the map to a printer or PDF file. Several export options exists for map documents, but only the export to PDF is described here. See the ArcGIS Help under **Geoprocessing—The ArcPy site package—Mapping module—Functions—Exporting and Printing Maps**.

To export or print the map the ExportToPDF or the PrintMap function is used. The script below shows the two different implementations. The Print Map functions are commented out so that the end user can change the settings depending on the local printer or network.

```
# Check to see if PDF exists, if it does, delete it
if arcpy.Exists(mappath + 'test_map.pdf'):
    arcpy.Delete_management(mappath + 'test_map.pdf')

print 'Writing PDF file...'

# Create the PDF document
ExportToPDF(mxd, mappath + 'test_map.pdf')
print 'Created : ' + 'test_map.pdf'

# Alternatively, the map can be printed to a local printer
# This is commented out and can be changed by the code developer

# PrintMap(mxd)    # prints map to default printer
# PrintMap(mxd, '\\\\network_location\\printer_name') # print to networked printer
```

Notice that the path of the map (the `mappath` variable) and a file name (`'test_map.pdf'`, "hard coded" in this example) is used in the `ExportToPDF` function to save the map to a PDF file on disk. The map document (the `mxd` variable above) will be exported to the path and file name in the second parameter of the `ExportToPDF` function. Note that the `mappath` variable is defined toward the top of the script. The top portion of the script is shown below.

```
author = 'N. Jennings'

# reformat date to MM.DD.YYYY
# NOTE: lowercase %y will result in MM.DD.YY format
# See Python.org or text regarding the datetime module

CUR_DATE = datetime.date.today().strftime('%m.%d.%Y')

try:

    # Get the map document.
    # In this case a custom template set up for map production
    # change the paths as needed to run the script form a local system
    datapath = 'C:\\PythonPrimer\\Chapter09\\'
    mappath = datapath + 'MyData\\Maps\\'    # output for PDFs
    mxd = MapDocument(datapath + 'Mapping_Module_Overview.mxd')
```

To print the map document to a local printer, the `PrintMap` function is used which refers to the map document as the parameter.

Saving Map Documents

ArcMap documents can be saved after changes have been made to them using Python by using either the save() or the saveACopy. The save() method performs the same operation as **File—Save** in ArcMap, while saveACopy() performs the same operation as **File—SaveACopy** in ArcMap. The save method will save any changes made with the Python script. The saveACopy method will save the changes to a new ArcMap path and file name (defined in the script) and an optional ArcGISversion value (e.g 9.3, 9.2, etc). The default is ArcGIS 10, but other previous versions can be used in the version parameter. The save method may not be desirable, if the changes made using the script are used for map document production where PDF files or hard copy prints are made. By not saving the changes to the map template used in the map production routine, the map remains in its original state. In the case of creating and executing a script for map production, the script makes changes to the map as needed and then uses the export or print routine to create the PDF files or print the maps to a printer, so using the save or saveAsCopy are really not required. The script below shows the save and saveACopy methods as being "commented out" so a programmer can decide to use them or not.

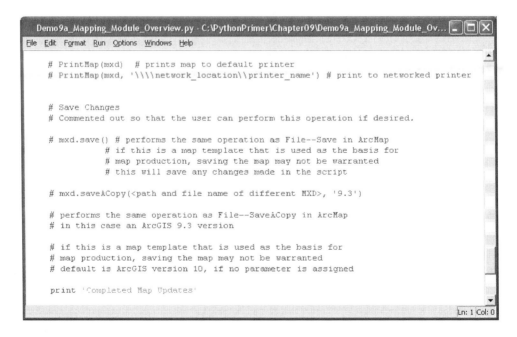

Working with Data Frame, Layer, and Layout Element Methods

So far this chapter has focused on the fundamental map elements and some of the properties associated with them. In addition to the various properties each of the map elements may have, several methods are also available for data frames, layers, and layout elements. The section above regarding Saving Map Documents discussed and illustrated two common methods associated with the map document (`save` and `saveACopy`). This section will focus on some of the common methods associated with data frames, layers, and layout elements such as changing the zoom extent.

When automating the map production process, changing the map extent of the data frame in Layout View is often required. A number of methods provide this capability and can be implemented in a number of ways depending on the information and level of detail needed in a given map. Changing the map extent involves applying the new extent to the data frame. The map extent can be changed by using a definition query or changing the extent based on selected features from all layers or from a single layer. These methods are described below.

Changing the Map Extent using a Definition Query (or not)

One common method to change the map extent is to use the `definitionQuery` Layer property to limit a specific area or geographic location based on a set of criteria (such as a specific project location or neighborhood).

```
query = '"NAME" = \'Newton Booth\''
TOCLayer.definitionQuery = query
```

This property can be useful to show a focus area where other geographic data provide some context (such as roads). The map below shows a neighborhood shown in a specific color (e.g. from a neighborhood layer) where the gray background and roads are shown in different colors and symbology. The specific neighborhood has been isolated as a result of a definition query. The neighborhood appears as the focus of the map.

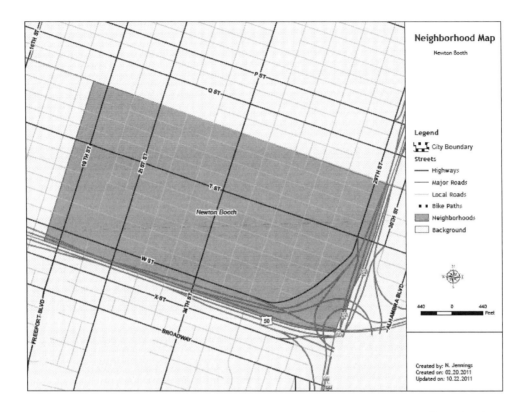

From a code development point of view the following syntax is used. A looping structure is used to cycle through all of the layers in the map document using a ListLayers routine as illustrated above. See the **Accessing Layers** section above. The TOCLayer is a variable that points to a specific layer (in this case the neighborhoods layer). A query is created to identify a specific neighborhood (*'Newton Booth'*). This query is applied to the definitionQuery property of the layer. The next line uses the getExtent() method (a method in the Layer class) of the layer and assigns it to the dataframe.extent property, which effectively changes the map extent to the extent of the "queried" features in the definition query. The getExtent() will get the extent of the feature (or features) that meet the criteria of the definition query. The dataframe.scale = dataframe.scale * 1.1 applies a ten percent buffer around the extent of the queried feature that prevents the edges of the feature from touching the map frame (i.e. it makes the map a little more pleasing to view). Applying this scale factor is common practice when creating map sets or map books.

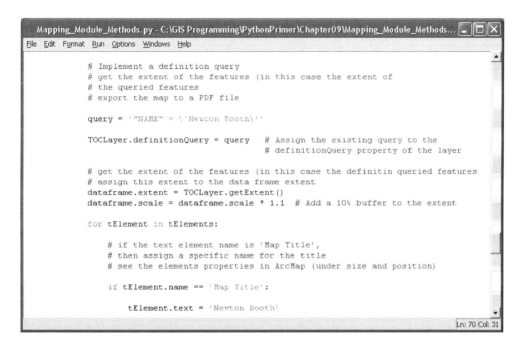

If a definition query is not used then the `getExtent()` method will return the full extent of the data layer, which the code developer may not desire.

Changing the Map Extent using Selected Features

Oftentimes, zooming to the extent of one or more selected features is required when developing automated processes for creating maps or map books. Automated map sets and books usually have a focus area (such as a project area, neighborhood, map grid, etc.). In addition, in more complex map books, multiple map sheets may be required to show the appropriate level of detail of geographic features, labels, and symbols. To take advantage of the ability to automate the map production workflow, being able to change the map extent to a selected feature or features is very useful. ArcGIS using Python provides a couple of methods that help make this possible.

1. `zoomToSelectedFeatures()`
2. `getSelectedExtent()`

The first method is in the data frame class and can change the data frame's extent based on any set of features that are selected in all layers. This functionality mimics the **Select—Zoom To Selected Features** option in ArcMap. The second method is in the layer class and can change the data frame's extent based on the selected feature within a specific layer. This functionality mimics the **Selection—Zoom To Select Features** option found under the context menu when the user right-clicks on a layer's name in the Table of Contents (i.e. the list that is often activated when the user needs to look at the layer's properties). Both methods can be used to change the map display and scale of the map to focus on a specific area. Either method is often preceded by a selection method (`SelectLayerByAttribute` or `SelectLayerByLocation`) to create a set of selected features that can then be used to change the display of the map.

The code below shows an example where the `zoomToSelectedFeatures()` **method is used.**

```python
# Neighborhoods is a layer within the group

if TOCLayer.longName == 'Detail Map\Neighborhoods':

    # perform a Select by Attriute to select a single neighborhood

    query = '"NAME" = \'Alkali Flat\''
    print query

    arcpy.SelectLayerByAttribute_management(TOCLayer, "NEW_SELECTION", query)

    result = arcpy.GetCount_management(TOCLayer)
    print 'Number of selected features: ' + str(result)

    # zoom to the extent of the selected feature

    dataframe.zoomToSelectedFeatures()

    #tElements = ListLayoutElements(mxd, "TEXT_ELEMENT")

    # cycle through the text elements to change the map title and date
    for tElement in tElements:

        # if the text element name is 'Map Title',
```

A loop is used to access a list of layers from the map document. The `TOCLayer` variable points to one of these layers. In the case above, it points to the **Neighborhoods** layer that is part of the **Detail Map** group layer. A query is defined that is used in the `SelectLayerByAttributes` routine. Once the feature is selected, the `zoomToSelectedFeatures()` method is implemented. Notice that it is used with a `dataframe` variable (defined towards the top of the script), since the `zoomToSelectedFeatures()` is a method of the Data Frame class.

```
dataframe.zoomToSelectedFeatures()
```

When the map is eventually exported to a PDF document, the resulting map shows the highlighted feature.

If other features from different layers were selected, the `zoomToSelectedFeatures()` method would zoom to the extent of all of the selected features.

The `getSelectedExtent()` method works in a similar manner, but is limited to a specific layer's selected features because its method is related to the layer (a method in the Layer class). The script below illustrates the `getSelectedExtent()` method.

```
# clear any selected features
arcpy.SelectLayerByAttribute_management(TOCLayer, "CLEAR_SELECTION")

dataframe.zoomToSelectedFeatures()

print 'Exporting Map 2...'
if arcpy.Exists(mappath + 'NoSelectedFeaturesZoom_Map2.pdf'):
    arcpy.Delete_management(mappath + 'NoSelectedFeaturesZoom_Map2.pdf')

ExportToPDF(mxd, mappath + 'NoSelectedFeaturesZoom_Map2.pdf')

query = '"NAME" = \'Downtown\''
print query

arcpy.SelectLayerByAttribute_management(TOCLayer, "NEW_SELECTION", query)

dataframe.extent = TOCLayer.getSelectedExtent()

#tElements = ListLayoutElements(mxd, "TEXT_ELEMENT")

# cycle through the text elements to change the map title and date
# do this again because the map changed neighborhoods
for tElement in tElements:
```

The script, as shown above, uses a query string in a `SelectLayerByAttributes` function to select a specific neighborhood from the neighborhoods layer (the same layer referred to above). After the selected features are obtained, the getSelectedExtent () method is used which is then assigned to the data frame's extent property.

```
dataframe.extent = TOCLayer.getSelectedExtent()
```

When the map is eventually exported to a PDF document, the extent of the map will represent that of the selected feature (in this case the selected neighborhood). The resulting map will show the feature highlighted.

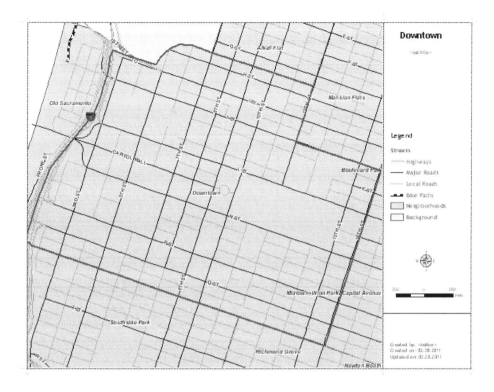

Notice above that the highlighted feature's edges touch the sides of the data frame. The script does not implement the `dataframe.scale *1.1` as previously shown. If it did, the map would be "zoomed out" slightly, where the feature's edges would not touch the data frame. If the programmer wanted to "clear" the selected features, the `SelectLayerByAttribute` routine could be used with the 'CLEAR_SELECTION' selection type before the map is exported or printed.

Adding and Saving Layer Files

The last two methods to discuss in this chapter are adding and saving layer (`.lyr`) files. A programmer may want to dynamically add and save layer files as part of the map production process. For example, a major roads layer may need to be added to certain kinds of map documents. If a programmer wants to "save" an existing layer file to a new name as part of the map production process, this can also be performed.

When datasets are added to the Table of Contents, feature classes automatically become feature layers. The layers in the Table of Contents often have the symbology changed as well showing labels and applying definition queries. The layer "conditions" can be saved to a layer file with the (.lyr) extension and can be added to use in other map documents. If a programmer wants to add an existing layer to a map programmatically, the AddLayer function can be used. See **Demo 9b, Map 7**.

```
# datapath - path to layer file
# dataframe - variable for a specific data

aLayer = Layer(datapath + 'aLayerFile.lyr')
AddLayer (dataframe, aLayer)
```

When an ArcGIS user saves a layer file in ArcMap, he/she typically performs this function by right-clicking on the layer name and choosing **Save As Layer File**. This same function can also be performed programmatically by using the save() or saveAsCopy() method. To overwrite an existing layer (.lyr) file, the save() method is used; to save the layer to a new (.lyr) file, the saveACopy() method is used. Both of these are related to the layer (i.e. found in the Layer class).

NOTE: From a Python programming point of view, specific changes to the symbology cannot be made (such as changing colors, labels, line types, etc); however, the Layer class does have properties to change the name, transparency, visibility, and labeling of the layer in the map document.

The following Python script snippet illustrates how a layer file can be added to a map document, have one of its properties (showLabels) changed, and then be saved to a new layer file.

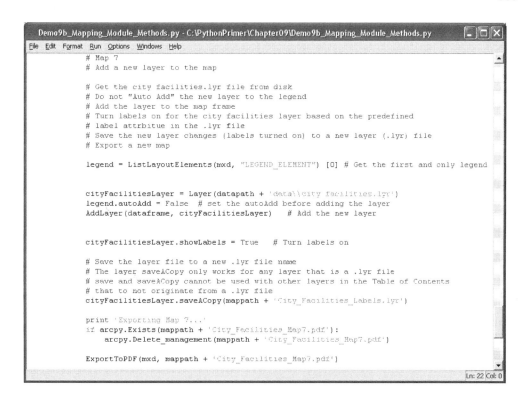

The script above is part of the **Demo 9b Mapping Module Methods** demonstration script that shows a variable (legend) being assigned to the first (and only) legend item found in the map document from a list of layout elements. The next line locates the .lyr file on disk and assigns this to a layer variable (cityFacilitiesLayer). The .lyr file contains settings such as symbology, label types and fonts, definition queries, etc. and are part of the original layer file. Before the layer is added, the legend's autoAdd property is set to False so that the new layer is not added to the map's legend. The layer is then added to the map's data frame using the AddLayer function. Since this map only has one data frame, which is defined towards the top of the code, the layer is added to this data frame. Next, the layer's labels are turned on using the showLabels property. Any label settings (fields, fonts, types, etc.) that are contained in the .lyr file definition and will be displayed for the layer. A new layer file can be created by using the saveACopy method and will include any definition queries, label changes, transparency, or brightness changes that were modified programmatically. The current layer definition can be overwritten by using the save() method. The map resulting from the changes above is exported to a PDF file, which looks like the map below.

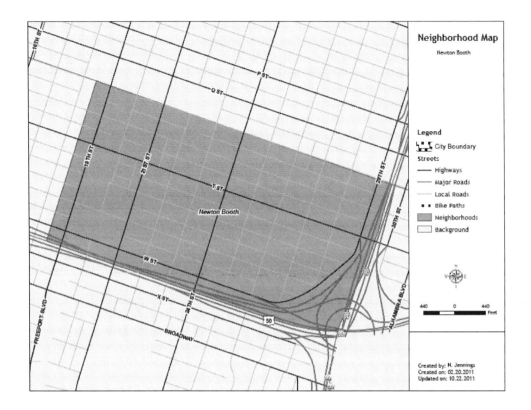

Creating a Map Book Programmatically using the `mapping` Module

One goal of creating maps programmatically can be to create a map book (or map set) that contains a number of different maps where the differences are typically the geographic data displayed in the data frame, title, subtitle, legend, scale, date, and other title block information such as a map sheet number or index. As mentioned previously, ArcGIS has a Data Driven set of tools to assist the creation of a map book, such as creating a map sheet index feature class and creating attribute fields that can be used in naming map sheets. These can be useful when developing a map production routine. This section will cover some of the fundamentals of programmatically creating a number of unique map sheets outside the use of the Data Driven Pages tools and routines. Additional map elements can be added and modified as the programmer becomes more proficient with the methods illustrated in this chapter.

If the reader closely reviews the **Demo 9b Mapping_Module_Methods.py** script he/she will note that multiple maps are created by using conditional statements (e.g `if` statements). Within the conditional statements queries are built and used with selection routines to change the content in the data frame, rename the title, and export the map to an output map name. If the reader used the **Sacramento_Neighborhoods.shp** file, there are 129 unique neighborhoods. Creating a map for each one using conditional statements could be quite cumbersome.

A more flexible method is to use a search cursor and a looping structure to iterate through each unique neighborhood in neighborhood shapefile and variables in query strings so that feature selections (such as `SelectLayerbyAttribute` or `SelectLayerByLocation`) or definition queries can be performed to display the desired geographic data in the map's data frame. Looping structures can also be used to iterate through text elements to modify titles, subtitles, map sheet names, etc. and for changing the display properties of layers in the table of contents. The reader will see some of these elements illustrated in the demonstration scripts for this chapter. The practical implementation of creating a series of map sheets is the goal of Exercise 9.

Summary

This chapter has covered many of the fundamental programming tasks that can be used to auto-generate map sheets or map books. It has introduced the primary requirements of accessing an ArcMap document, its layers, data frame, and map elements. In addition, this chapter has covered many of the specific methods and properties that are often used when creating maps and map books. This chapter serves as a launch point to add more elements and functionality to the map automation process. For example, if the code developer needed to use a separate map grid system to create multiple map sheets that maintained a single scale, an additional layer can be used that contains these specific map grid shapes as well as attributes that contain the unique map sheet name. The code would essentially have a more involved looping strategy (possibly multiple loops) to generate all of the maps for the map series. The fundamentals remain the same, just a bit more complex to implement (and can end up saving a lot of time creating unique maps with relatively minor changes).

Demos Chapter 9

The demonstrations in this chapter are organized into two scripts:

1. the mapping module overview (**Demo 9a**), which focuses on the script organization to access different parts of a map, such as the map document, data frame, layers, and layout elements as well as manipulating some of the respective properties

2. the mapping module methods (**Demo 9b**), which focus on changing the data frame's extent through the use of selected features and definition queries and making changes to the layers and layout elements.

A series of seven maps are produced from the **Demo 9b** script.

Demo 9a: Mapping Module Overview and Properties

This script demonstrates some of the fundamentals of the map document, layers, and layout elements. See the **\PythonPrimer\Chapter09\Demo9a_Mapping_Module_Overview.py** script file and the **Mapping_Module_Overview.mxd** map document. Make sure to change the data paths appropriately to properly run the script.

A unique element in this script is the use of the `from arcpy.mapping import *` line. This line "imports the entire mapping module" so that some of the code related to the mapping module can be simplified such that one need not type `arcpy.mapping` in front of the entire mapping module related functions, properties, and methods. See the Chapter 9 text for more details or consult the ArcGIS Help documents.

The reader can use this demo to help develop the programming code for creating an automated map routine as well as be able to change some of the properties to see the effect on the resulting map (**test_map.pdf**) that is created at the end of the script.

After setting up the data paths pointing to the ArcMap document, the `MapDocument` function is used to locate the map document on disk. A pre-existing map document must exist before any other mapping functions can be implemented.

```
import arcpy, sys, traceback, datetime
from arcpy.mapping import *    # Eliminates the need to write arcpy.mapping
                               # before each mapping module related class, function
                               # or method

author = 'N. Jennings'

# reformat date to MM.DD.YYYY
# NOTE: lowercase %y will result in MM.DD.YY format
# See Python.org or text regarding the datetime module

CUR_DATE = datetime.date.today().strftime('%m.%d.%Y')

try:

    # Get the map document.
    # In this case a custom template set up for map production
    # change the paths as needed to run the script form a local system
    datapath = 'C:\\PythonPrimer\\Chapter09\\'
    mappath = datapath + 'MyData\\Maps\\'      # output for PDFs
    mxd = MapDocument(datapath + 'Mapping_Module_Overview.mxd')

    # Report some of the Map Document Properties

    print 'Map Document Title: ' + str(mxd.title)
    print 'Map Document Author: ' + str(mxd.author)
    print 'Map Document Summary: ' + str(mxd.summary)
    print 'Map Document Desription: ' + str(mxd.description)
    print 'Map Document last Date Saved: ' + str(mxd.dateSaved)
    print 'Is the Map Document Relative Path checked? : ' + str(mxd.relativePaths)

    # Get a list of data frames
    # [0] inicates the first (or only data frame, in this case)
    # The data frame name is "Layers", the default name for the
    # primary data frame
    dataframe = ListDataFrames(mxd, "Layers") [0]

    # Report some properties about the map frame named "Layers"
```

Next a series of map document properties are reported to the Python Shell so that the reader can see that a number of different properties exist for the map document. All of these properties can be found under the Map Document Properties on the File menu in ArcMap.

The next step when working with the map document is to obtain a data frame. ArcGIS uses a `ListDataFrames` function to get a list of data frames. Most of the time only one exists, but a given map page can have more than one data frame. ArcGIS uses a Python list to keep track of the data frames. To access a specific data frame, the `[index]` value can be used. The index value refers to the position in the Python list the data frame is stored. Python lists have a starting index value of zero (or the first element in the list). As shown above, the first data frame is assigned to the `dataframe` variable.

```
# primary data frame
dataframe = ListDataFrames(mxd, "Layers")[0]

# Report some properties about the map frame named "Layers"

print 'Map Frame Map Units: ' + str(dataframe.mapUnits)
print 'Map Frame Scale: ' + str(dataframe.scale)

# See ArcGIS Help under 'spatialreference' (all one word) for more information
print 'Map Spatial Reference: ' + str(dataframe.spatialReference.name)
print 'Map Frame Anchor Point X Position (page units): ' + str(dataframe.element
print 'Map Frame Anchor Point Y Position (page units): ' + str(dataframe.element
print 'Map Frame Width (page units): ' + str(dataframe.elementWidth)
print 'Map Frame Height (page units): ' + str(dataframe.elementHeight)

# Get the extent of the data frame and assign it to a variable
mapExtent = dataframe.extent

# The following syntax prints out the Min and Max X and Y values
# for the map extent of the data frame.
# Search on Extent in the ArcGIS Help to find more details on
# the extents XMin, YMin, XMax, and YMax

# The string formatting can be found in a Python text or at Python.org
# %f indicates a decimal value will be added to a string
# each %f below is substituted with the appropriate mapExtent.<value>

print 'Map Extent of Data Frame is: \n' + \
    'XMin: %f, YMin: %f, \nXMax: %f, YMax: %f' % \
        (mapExtent.XMin, mapExtent.YMin, mapExtent.XMax, mapExtent.YMax)

# Get a list of layers in the table of contents of the map document
TOCLayers = ListLayers(mxd)

print 'Processing layout elements...'

# loop through the layers
for TOCLayer in TOCLayers:
    print 'Layer Name: ' + str(TOCLayer.name)
```

A series of `print` statements are written to the Python Shell again so that the user can see a number of the properties available for the data frame. The X and Y anchor points as well as the width and height can be used to position the data frame on the map layout page. These values are provided in page units and so often refer to inches on a page with the lower left corner of a page having X and Y values of (0,0). Also shown is the use of the data frame's extent where the next line provides a report of the specific X and Y minimum and maximum values. These might be needed to set other extents in the map layout or change the display of the geographic data.

The next section of the script illustrates how to access a layer. Similar to the data frame, a list (`ListLayers`) is used to obtain a list of layers in the map document. Like the data frame, a single layer could have been selected by obtaining its position in the table of contents. Often a list is retrieved for all layers and then a `for` loop, like the one shown below, is used to iterate through the list. Then, if a specific set of processes is required on a specific layer (such as turning it off, changing transparency, labeling, etc.) `if` statements are used. In this example, a number of layer properties are shown. One in particular is the `longName` property. The `longName` property can be used to access or check if a layer is a group layer and has layers within it.

```
# Get a list of layers in the table of contents of the map document
TOCLayers = ListLayers(mxd)

print 'Processing layout elements...'

# loop through the layers
for TOCLayer in TOCLayers:
    print 'Layer Name: ' + str(TOCLayer.name)

    print 'Longname: ' + str(TOCLayer.longName)   # longName provides
                                                  # the ability to use
                                                  # "Group Layers"

    # Detail Map is the group layer
    # Neighborhoods is a layer within the group

    if TOCLayer.longName == 'Detail Map\Neighborhoods':

        TOCLayer.transparency = 50   # 50% transparency

    if TOCLayer.longName == 'Detail Map\Streets':

        TOCLayer.showLabels = True   # turn labels on

    if TOCLayer.longName == 'Detail Map\Parcels':

        TOCLayer.visibility = False   # turn layer off

# Get a list of "text" type layout elements from the map document
tElements = ListLayoutElements(mxd, "TEXT_ELEMENT")

print 'Processing text elements...'
for tElement in tElements:

    # if the text element name is 'Map Title',
    # then assign a specific name for the title
    # see the elements properties in ArcMap (under size and position)
```

The script continues by creating a list of layout elements, in this case "text elements." If no layer element type is used, such as a `TEXT_ELEMENT`, then all layout elements can be obtained. See Chapter 9 and the ArcGIS help for more specifics. Remember that the map document contains uniquely named layout elements that are named in the existing map document so that a list like this can programmatically access and change the properties of the layout elements as required. Like the data frame and layers, a `for` loop can be used to iterate through each element and make changes.

```
# Get a list of "text" type layout elements from the map document
tElements = ListLayoutElements(mxd, "TEXT_ELEMENT")

print 'Processing text elements...'
for tElement in tElements:

    # if the text element name is 'Map Title',
    # then assign a specific name for the title
    # see the elements properties in ArcMap (under size and position)

    if tElement.name == 'Map Title':

        tElement.text = 'Test Map'
        tElement.elementPositionX = 8.7  # Anchor point is lower left
                                         # The anchor point is used
                                         # to locate the lower left
                                         # page position of the text

        # The X and Y positions (and Height and Width
        # can be found in the element properties
        # of the layout object under Size and Position

    if tElement.name == 'Print Date':

        tElement.text = str(CUR_DATE)  # assign the current date
                                       # to this text element


# Check to see if PDF exists, if it does, delete it
if arcpy.Exists(mappath + 'test_map.pdf'):
    arcpy.Delete_management(mappath + 'test_map.pdf')

print 'Writing PDF file...'

# Create the PDF document
ExportToPDF(mxd, mappath + 'test_map.pdf')
print 'Created : ' + 'test_map.pdf'
```

The script above shows some of the common properties that are changed with layout elements. For text elements, the text displayed on the map can be changed as well as the position (using `elementPositionX` and/or `elementPositionY`, the `elementWidth`, and the `elementHeight`), similar to the data frame position properties. To set the date of the map, the current date of the computer system is used. Toward the top of the script, the variable `CUR_DATE` is set to the current system data in a format that indicates MM.DD.YYYY. Consult a Python text or the *Python.org* site for additional date format options. The `datetime` module has also been imported to provide the functionality of dates and times.

Once all of the changes have been made, the map can be printed or exported to one of the supported formats. One of the most common export formats is PDF. The user must be have a PDF reader or viewer to open the resulting PDF. The script below shows the `ExportToPDF` function to generate the PDF file.

The commented sections of the script also shows the use of `PrintMap`, `save`, and `saveACopy` routines. These can be uncommented and tested on the user's system. The `save` routine performs the same function as **File—Save**, while `saveACopy` performs the same function as **File—SaveACopy** from the ArcMap File menu. `saveACopy` can be used to save the changes made to the ArcMap document to a new file or even to a previous ArcGIS version. `save` will overwrite any changes made by the script to the existing map document.

```
    # Check to see if PDF exists, if it does, delete it
    if arcpy.Exists(mappath + 'test_map.pdf'):
        arcpy.Delete_management(mappath + 'test_map.pdf')

    print 'Writing PDF file...'

    # Create the PDF document
    ExportToPDF(mxd, mappath + 'test_map.pdf')
    print 'Created : ' + 'test_map.pdf'

    # Alternatively, the map can be printed to a local printer
    # This is commented out and can be changed by the code developer

    # PrintMap(mxd)   # prints map to default printer
    # PrintMap(mxd, '\\\\network_location\\printer_name') # print to networked print

    # Save Changes
    # Commented out so that the user can perform this operation if desired.

    # mxd.save() # performs the same operation as File--Save in ArcMap
             # if this is a map template that is used as the basis for
             # map production, saving the map may not be warranted
             # this will save any changes made in the script

    # mxd.saveACopy(<path and file name of different MXD>, '9.3')

    # performs the same operation as File--SaveACopy in ArcMap
    # in this case an ArcGIS 9.3 version

    # if this is a map template that is used as the basis for
    # map production, saving the map may not be warranted
    # default is ArcGIS version 10, if no parameter is assigned

    print 'Completed Map Updates'
except:
```

Demo 9b: Implementing Mapping Module Methods

The previous demonstration focused on working through the logical organization of changing different map elements on a map page. This demonstration script focuses on the methods (or the actions) that can be used to change the map page, primarily the geographic data in the data frame. A number of different maps are created to illustrate different ways to change the geographic data in the map frame (as well as some of the layout elements). The pre-created maps (PDF documents) can be found under
\PythonPrimer\Chapter09\MyData\Maps. Refer to
Demo9b_Mapping_Module_Methods.py and the
Mapping_Module_Overview.mxd for a completed script and the associated map document used in this demo.

The first part of the demonstration script is very similar to **Demo 9a**. A map document, the data frame, and layers are referenced. When the loop structure is used to iterate over the layers a number of `if` statements are written to produce the different maps. This is not a real efficient way of creating the different maps, but for demonstration purposes, this method works fine.

Map 1

The code shown below begins by creating and using a query to select a specific neighborhood from the neighborhood layer. After reporting the count of selected features, the `zoomtoSelectedFeatures()` method is used to modify the data frame to display the extent of the selected features. The `zoomToSelectedFeatures()` method zooms to the extent of all selected features in all layers in the data frame. In this case, since only a single feature is selected from a single layer, the data frame zooms to the extent of a single selected feature in the neighborhood layer.

```
print 'Processing layout elements...'

# loop through the layers
for TOCLayer in TOCLayers:
    #print 'Layer Name: ' + str(TOCLayer.name)

    #print 'Longname: ' + str(TOCLayer.longName)   # longName provides
                                                   # the ability to use
                                                   # "Group Layers"

    # Detail Map is the group layer
    # Neighborhoods is a layer within the group

    if TOCLayer.longName == 'Detail Map\Neighborhoods':

        # perform a Select by Attriute to select a single neighborhood

        query = '"NAME" = \'Alkali Flat\''
        print query

        arcpy.SelectLayerByAttribute_management(TOCLayer, "NEW_SELECTION", query)

        result = arcpy.GetCount_management(TOCLayer)
        print 'Number of selected features: ' + str(result)

        # zoom to the extent of the selected feature

        dataframe.zoomToSelectedFeatures()

        #tElements = ListLayoutElements(mxd, "TEXT_ELEMENT")

        # cycle through the text elements to change the map title and date
        for tElement in tElements:

            # if the text element name is 'Map Title',
            # then assign a specific name for the title
            # see the elements properties in ArcMap (under size and position)
```

Next, several layout properties are updated through the use of the `ListLayoutElements` routine and a `for` loop to update specific text element properties and then the `ExportToPDF` routine is used to generate an output PDF. See below.

```
# cycle through the text elements to change the map title and date
for tElement in tElements:

    # if the text element name is 'Map Title',
    # then assign a specific name for the title
    # see the elements properties in ArcMap (under size and position)

    if tElement.name == 'Map Title':

        tElement.text = 'Alkali Flat'

    if tElement.name == 'Print Date':

        tElement.text = str(CUR_DATE)   # assign the current date
                                        # to this text element
print 'Exporting Map 1...'
if arcpy.Exists(mappath + 'ZoomToSelectedFeatures_Map1.pdf'):
    arcpy.Delete_management(mappath + 'ZoomToSelectedFeatures_Map1.pdf')

ExportToPDF(mxd, mappath + 'ZoomToSelectedFeatures_Map1.pdf')

# Map 2
# clear any selected features
arcpy.SelectLayerByAttribute_management(TOCLayer, "CLEAR_SELECTION")


dataframe.zoomToSelectedFeatures()

print 'Exporting Map 2...'
if arcpy.Exists(mappath + 'NoSelectedFeaturesZoom_Map2.pdf'):
    arcpy.Delete_management(mappath + 'NoSelectedFeaturesZoom_Map2.pdf')

ExportToPDF(mxd, mappath + 'NoSelectedFeaturesZoom_Map2.pdf')

# Map 3
```

Map 2

Map 2 shows the map display if no features are selected when implementing the `zoomToSelectedFeatures()` method. In this case all features are cleared and the map zooms to the full extent of the data. See above.

Map 3

For Map 3 the query is changed to select a different neighborhood. A feature selection is performed using the `SelectLayerByAttribute` function. The layer's extent is retrieved from by using the `getSelectedExtent()` method that is related to the layer. This extent is then used to set the data frame's extent. The rest of the code changes a couple of text layout elements and exports the map.

```python
        print 'Exporting Map 2...'
        if arcpy.Exists(mappath + 'NoSelectedFeaturesZoom_Map2.pdf'):
            arcpy.Delete_management(mappath + 'NoSelectedFeaturesZoom_Map2.pdf')

        ExportToPDF(mxd, mappath + 'NoSelectedFeaturesZoom_Map2.pdf')

        # Map 3
        query = '"NAME" = \'Downtown\''
        print query

        arcpy.SelectLayerByAttribute_management(TOCLayer, "NEW_SELECTION", query)

        dataframe.extent = TOCLayer.getSelectedExtent()

        #tElements = ListLayoutElements(mxd, "TEXT_ELEMENT")

        # cycle through the text elements to change the map title and date
        # do this again because the map changed neighborhoods
        for tElement in tElements:

            # if the text element name is 'Map Title',
            # then assign a specific name for the title
            # see the elements properties in ArcMap (under size and position)

            if tElement.name == 'Map Title':

                tElement.text = 'Downtown'

        print 'Exporting Map 3...'
        if arcpy.Exists(mappath + 'GetSelectedExtentfromLayer_Map3.pdf'):
            arcpy.Delete_management(mappath + 'GetSelectedExtentfromLayer_Map3.pdf')

        ExportToPDF(mxd, mappath + 'GetSelectedExtentfromLayer_Map3.pdf')
```

Map 4

The reader will see that the neighborhood in Map 3 shows up as "highlighted" as a result of the selection. Before Map 4 is produced the selection is cleared and the map exported. Map 4 shows the same area, but without the highlighted feature.

```
for tElement in tElements:

    # if the text element name is 'Map Title',
    # then assign a specific name for the title
    # see the elements properties in ArcMap (under size and position)

    if tElement.name == 'Map Title':

        tElement.text = 'Downtown'

print 'Exporting Map 3...'
if arcpy.Exists(mappath + 'GetSelectedExtentfromLayer_Map3.pdf'):
    arcpy.Delete_management(mappath + 'GetSelectedExtentfromLayer_Map3.pdf')

ExportToPDF(mxd, mappath + 'GetSelectedExtentfromLayer_Map3.pdf')

# Clear feature selection to remove the "highlighted features"

# Map 4
# clear any selected features, but don't change the zoom extent
arcpy.SelectLayerByAttribute_management(TOCLayer, "CLEAR_SELECTION")

print 'Exporting Map 4...'
if arcpy.Exists(mappath + 'GetSelectedExtentfromLayer_NoHighlight_Map4.pdf'):
    arcpy.Delete_management(mappath + 'GetSelectedExtentfromLayer_NoHighlight_Map4.pdf')

ExportToPDF(mxd, mappath + 'GetSelectedExtentfromLayer_NoHighlight_Map4.pdf')

# Map 5
```

Map 5

The code to create Map 5 shows the same query used for Map 4, but this time the query is used as the `definitionQuery` property for the layer. The definition query will display only the features that meet the criteria, in this case, a specific neighborhood. In addition to using the definition query property, the data frame's scale is also changed slightly to zoom out by 10% so that the feature's edges do not touch the map frame, making the map look a little more centered.

```
        # Map 4
        # clear any selected features, but don't change the zoom extent
        arcpy.SelectLayerByAttribute_management(TOCLayer, "CLEAR_SELECTION")

        print 'Exporting Map 4...'
        if arcpy.Exists(mappath + 'GetSelectedExtentfromLayer_NoHighlight_Map4.pdf'):
            arcpy.Delete_management(mappath + 'GetSelectedExtentfromLayer_NoHighlight_Map4.pdf')

        ExportToPDF(mxd, mappath + 'GetSelectedExtentfromLayer_NoHighlight_Map4.pdf')

        # Map 5
        TOCLayer.definitionQuery = query      # Assign the existing query to the
                                              # definitionQuery property of the layer


        dataframe.scale = dataframe.scale * 1.1    # Add a 10% "buffer" to the extent
                                                   # so that the edges of the feature
                                                   # are not touching the map frame

        # The exported map will contain the definition query
        # the modified data frame scale
        print 'Exporting Map 5...'
        if arcpy.Exists(mappath + 'DefinitionQuery_ScaleChange_Map5.pdf'):
            arcpy.Delete_management(mappath + 'DefinitionQuery_ScaleChange_Map5.pdf')

        ExportToPDF(mxd, mappath + 'DefinitionQuery_ScaleChange_Map5.pdf')
```

Map 6

Map 6 is similar to Map 3, but uses the `getExtent()` method for the neighborhood layer that has a definition query applied to it. Again, the data frame's scale is zoomed out by 10% so that the feature's edges do not touch the data frame. Some of text elements are changed and the map is exported.

```python
# Map 6
query = '"NAME" = \'Newton Booth\''

TOCLayer.definitionQuery = query   # Assign the existing query to the
                                   # definitionQuery property of the layer

# get the extent of the features (in this case the definitin queried features
# assign this extent to the data frame extent
dataframe.extent = TOCLayer.getExtent()
dataframe.scale = dataframe.scale * 1.1  # Add a 10% buffer to the extent

for tElement in tElements:

    # if the text element name is 'Map Title',
    # then assign a specific name for the title
    # see the elements properties in ArcMap (under size and position)

    if tElement.name == 'Map Title':

        tElement.text = 'Newton Booth'

# The exported map will contain the definition query
# the modified data frame scale
print 'Exporting Map 6...'
if arcpy.Exists(mappath + 'DefinitionQuery_GetExtent_ScaleChange_Map6.pdf'):
    arcpy.Delete_management(mappath + 'DefinitionQuery_GetExtent_ScaleChange_Map6.pdf')

ExportToPDF(mxd, mappath + 'DefinitionQuery_GetExtent_ScaleChange_Map6.pdf')
```

Map 7

Map 7 illustrates how to add a `.lyr` file to an existing data frame in addition to "not" changing the map's legend by using the `autoAdd = False` option for the existing legend (which uses the `ListLegendEelments` function to gain access to the map's legend. Also, in this map, the newly added layer's labels are turned on and then the modified layer file is saved to a new `.lyr` file using the `saveACopy()` method. The map that is exported shows the newly added layer with the labels visible.

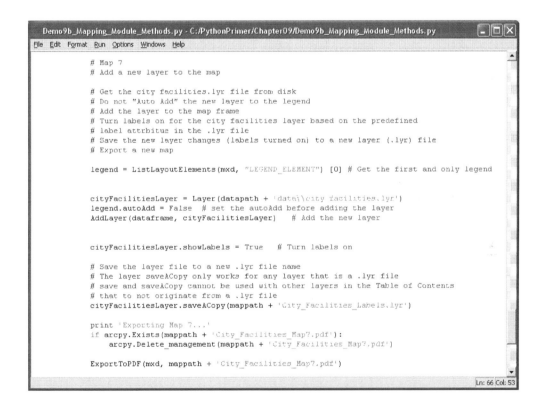

Exercise 9: Create a Simple Neighborhood Map Set

This exercise will expand on the concepts learned in this chapter where the code developer will automatically make a unique map for each neighborhood in the sample neighborhood set provided. The output maps will be exported to a PDF format. The reader should choose a folder to place the PDF files and use this in the script.

An ArcMap document (**Map_Template.mxd**) has been provided that contains the primary layers for the map as well as some of the layout elements.

A script called **Exercise9_StumpCode.py** can be used to work on the program. An outline is provided within the code to help the reader put additional code in the script. The steps are in the correct order. Use the demonstration scripts to assist with completing this exercise. All of the required elements can be found here. Some changes to the variables and syntax may need to be made depending on how the code developer names variables, etc.

Basically, the same kinds of steps will be used that are found within the demonstration scripts, however, a search cursor and a loop will be used to iterate over all of the neighborhoods in the neighborhood layer to update the required layout elements. An export function will be used to ouput each unique map in PDF format.

Notes

Also before getting started the reader should review the map document and the layer and layer elements to determine layer names, layer element names, and positions on the map. Also, it will be a good idea to review the neighborhood layer (**neighborhoods.shp**) and take a look at the names of the neighborhoods. **Sacramento_neighborhoos.shp** is used in map as the "Background" layer. It is NOT used in the script.

Indicators within the stump code script show where code needs to be indented so that loops can be properly implemented.

Change the data and map paths as necessary to locate the map and data as required.

The output should represent a uniquely named map that shows the specific neighborhood boundary centered in the data frame.

It will also be helpful to look at some of the questions below and answer them before beginning on the script.

Extra

A couple of "extra" sections can be found within the stump code. These are not required to implement the code. One section implements an `AddLayer` routine to add an existing layer file to the map and add it to the table of contents. A separate section can be implemented to reposition the neighborhood *'Midtown_Winn Park_Capital Avenue'* which is located within the *"Detail Map\Neighborhoods"* group layer. *'Midtown_Winn Park_Capital Avenue'* is a specific neighborhood within the *Neighborhoods* layer. See the attributes for the neighborhoods layer and the **Map_Template.mxd** Table of Contents as needed to develop the script.

Chapter 9: Questions

Answer the following based on information from the chapter. Use the ArcGIS Help documentation to supplement the information contained in Chapter 9.

1. What are the series of programming steps that are required to access a specific layer from a map?

2. Name 4 types of Layout Elements used in an ArcMap.

3. What are the specific element names for the following that can be found in the **Map_Template.mxd** document. The reader will likely need to open the map document and find the properties of the specific elements.

    a. Title
    b. Subtitle (where the specific neighborhood name will go)
    c. Author
    d. Date
    e. Legend

4. What kind of information is required to programmatically modify, export, or print map documents?

5. Describe what the following do and list two examples of each.

    a. Map Property

    b. Map Method

    c. Map Function

6. What Python routine is used to access a data frame?

7. What Python routine is used to access a layer?

8. What Python routine is used to access a layout element?

9. What Python routine is used to export a map to a PDF?

10. What Python routine is used to print a map to a local printer?

11. How can a map document that has been programmatically modified be saved?

12. What is the difference between `save()` and `saveACopy()` for the following:

    a. A Map document

    b. A layer file

13. What is the difference between the following?

    a. `getExtent`

    b. `getSelectedExtent`

    c. `zoomToSelectedFeatures`

14. What is the benefit of using a search cursor and this syntax for the query in the Exercise 9 script?

    `query = '"NAME" = \'' + NHName + '\''`

    where `NHName` represents a variable

15. In the Exercise 9 script (once you are able to export PDF maps), why does the **Map_Template.mxd** document NOT need to be saved?

Section III: Integrating and Automating Python Scripts for ArcGIS

Section III expands the capabilities already seen in the book by being able to use the Python script from an ArcGIS tool and to automatically execute a Python script in a batch process. Chapter 10 focuses on developing the custom tool interface, associating the Python script with an ArcGIS tool, developing parameters (values that are required to successfully execute the script) for the tool, and creating help documentation for the custom script tool. In addition, some minor modifications are required to pass parameters from the tool to the Python script. Chapter 11 illustrates the development of the batch routine which can be scheduled in a Windows scheduler to automatically execute the script at a specific date and time with a specific frequency (e.g. daily, weekly, monthly, etc.).

Chapter 10   Custom ArcGIS Tools and Python Scripts

Overview

Up to this point the scripts that have been written have been run (or started) from within the Python IDLE environment. Since *A Python Primer for ArcGIS* focuses on developing and using Python scripts for geoprocessing tasks, one might ask if Python scripts can be run or initiated from within the ArcGIS application. The short answer is yes. The programmer can create a custom toolbox to store a custom graphical user interface (GUI) that points to a Python script. An ArcToolbox can run an Esri function (often developed in C++ or similar programming language), a model developed from within ModelBuilder, and/or a custom Python script. Some of the existing ArcToolbox tools are developed in Python where the source code is exposed to the ArcGIS user. A common example referred to by Esri that uses Python is the **Multiple Ring Buffer** routine found within the **Analysis Toolbar—Proximity** Toolset. The tool interface and an excerpt from the script are shown below. The Python script for the **Multiple Ring Buffer** tool can be viewed by right-clicking on the tool and choosing Edit.

```
Version:     ArcGIS 9.4
Author:      Environmental Systems Research Institute Inc.
Required Arguments:
             An input feature class or feature layer
             An output feature class
             A set of distances (multiple set of double values)
Optional Arguments:
             The name of the field to contain the distance values (default="dis
             Option to have the output dissolved (default="ALL")
Description: Creates a set of buffers for the set of input features.  The buffer
             are defined using a set of variable distances.  The resulting featu
             class has the merged buffer polygons with or without overlapping
             polygons maintained as seperate features.
"""

import arcgisscripting
import os
import sys
import types

gp = arcgisscripting.create(9.3)

#Define message constants so they may be translated easily
msgBuffRings  = gp.GetIDMessage(86149) #"Buffering distance "
msgMergeRings = gp.GetIDMessage(86150) #"Merging rings..."
msgDissolve   = gp.GetIDMessage(86151) #"Dissolving overlapping boundaries..."

def initiateMultiBuffer():

    # Get the input argument values
    # Input FC
    input         = gp.GetParameterAsText(0)
    # Output FC
    output        = gp.GetParameterAsText(1)
    # Distances
    distances     = gp.GetParameter(2)
    # Unit
    unit          = gp.GetParameterAsText(3)
    if unit.lower() == "default":
```

Source: Esri, 2011. MultiRingBuffer Tool.

Providing a custom tool to an ArcGIS user provides a convenient method to "deploy" a script since the end user may not understand how a Python script functions. To create a custom tool that uses a Python script often a custom toolbox is created and then a script file is associated with it. The following section reviews the general workflow for creating a custom toolbox and associating a Python script with it.

Creating a Custom ArcToolbox

The basic steps involved to create a custom tool that uses a Python script are:

1. Create a custom toolbox and "Add" a script tool
2. Associate the Python script to the tool
3. Add and configure the parameters the script will need to execute
4. Modify the Python script to accept the tool parameter values
5. Create help documentation for the tool

Creating a custom ArcToolbox can be performed by right clicking on the ArcToolbox.

Browse to the location to place the new toolbox.

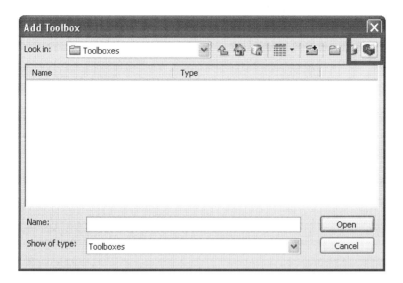

Click the "New Toolbox button above and change the name accordingly.

The new toolbox is added to the ArcToolbox.

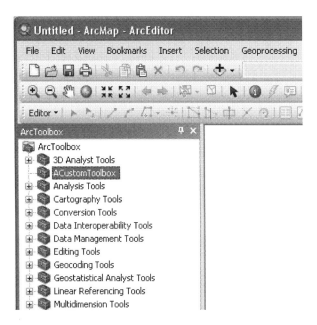

Note that if the new toolbox is located in a user's or networked folder that any new ArcMap document that is created will *not* contain this toolbox. To have the custom toolbox show up by default with all of the other standard toolboxes, the custom toolbox must be placed in the folder shown below on Windows XP machines. Windows Vista and Windows 7 may have a different path for the default ArcToolboxes.

```
C:\Program Files\ArcGIS\Desktop10.0\ArcToolbox\Toolboxes
```

Associating a Python Script to the Custom Toolbox

After the custom toolbox (e.g. **ACustomToolbox.tbx**) is created, the Python script (**ClipandBuffer.py**) can be added to it by right-clicking on the toolbox and choosing **Add** from the menu. A dialog box similar to the figure below will appear. Refer to the Chapter 10 material provided for this chapter and the demonstration script, **\Chapter10**.

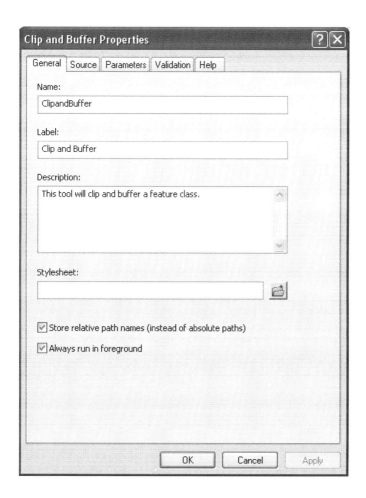

The General tab in the dialog box asks for a script name and Label. The script name is just a name and "cannot" contain spaces, dashes, or underscores. The name assigned, should be reasonable for the specific tool. The label is the name that will show up in the custom toolbox (and hence becomes the "script tool." The user can add a brief useful description about the functionality of the script. This description shows up in the tool help for the specific tool. It is recommended to always check the box to store relative path names.

Custom tools are also always run in the foreground by default. Tools that use foreground processing will commence execute until the tool completes. Foreground processing will prevent the use from using the application (such as ArcMap) until the tool completes. In addition, a progress dialog box will appear showing the progress of the tool. The reader is encouraged to read the **ArcGIS Help** topic **Geoprocessing—Executing tools—Foreground and background**

processing for more information. For *A Python Primer for ArcGIS*, this property will remain "checked" and the tool will be executing using foreground processing.

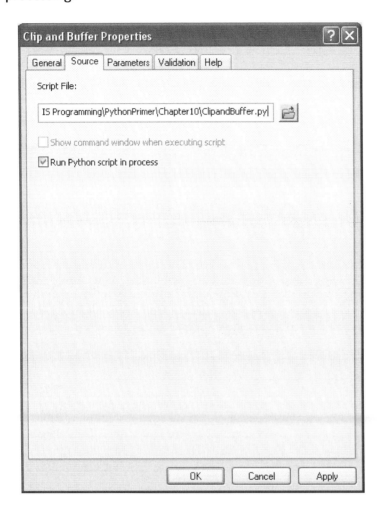

The Source tab in the dialog box requires the full path and file name to the associated Python script. It is recommended that the script be placed in some logical path so that it can be easily found so that additional modifications can be easily performed if necessary.

NOTE: If the user moves the custom toolbox to a different location, the Python script does not move with it. The Python script will remain in its location until the user changes this dialog box to associate the tool to the new location.

Defining Parameters for the Script Tool

Script tool parameters are created so the user of the tool can easily find and set the parameter values that will then be used by the script when the OK button is clicked. A good tool interface will include parameter names that clearly identity the kind of parameter needed by the script. In addition, the proper data type and some parameter properties can help limit the kinds of data or values that a user is able to enter. For example, if a code developer created a string data type for the buffer units, but did not limit the choices to a specific list of values that contain *'Feet'*, *'Meters'*, or *'Miles'*), the tool's user might type in the value *'FT'* instead of a value from the list. The value *'FT'* may not be recognizable for the specific ArcGIS tool that requires it, so the program will not function properly. To create a well-designed tool interface, the following are required:

1. Create a Display Name (the parameter label)
2. Set the proper Data Type for the parameter
3. Set Parameter Properties as required

Note: The tool properties can be viewed and modified by right-clicking on the tool name and selecting Properties after the script has been associated with the respective toolbox. If a script does not have any parameters defined, a blank tool interface is loaded and looks like the following figure.

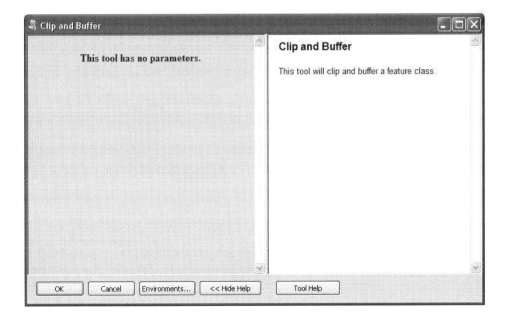

In addition, the Parameters Tab will also be blank.

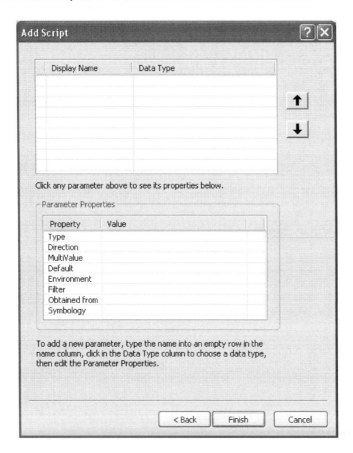

The following figure shows the parameters created for the custom **Clip and Buffer** tool when the Parameters Tab is displayed.

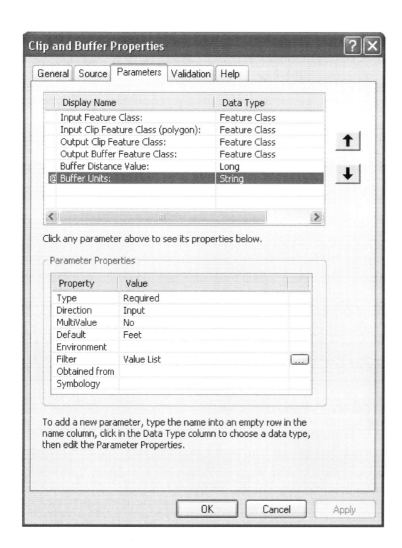

Notice the Display Name and Data Type columns at the top of the dialog box. The Display Name is the value that will show up as a parameter heading. The Data Type shows the type of value that the parameter will represent. The data types can be challenging depending on how "flexible" the code developer decides to make the script. For example, an initial development of a tool user interface may only limit user input to shapefiles, so a shapefile data type will likely be used for some of the parameters. If, on the other hand, the code developer wants the user to be able to choose a shapefile or a geodatabase feature class, then the feature class data type might be used. Also notice the Parameter Properties in the middle area of the dialog box includes some additional parameter setting that may be modified to limit how the parameters are used and can enforce the selection of appropriate values for the parameters. See the ArcGIS Help document under **Geoprocessing—Creating Tools—Creating script tools with Python scripts—Setting script tool parameters**.

Parameter Display Name

The display name is simply a short text phrase that will be used as the parameters label and indicates the kind of information required for the parameter. For example, a Display Name might be assigned to "Input Polygon Feature Class:" to indicate to the user that a polygon feature class should be chosen. The actual Data Type and parameters set up by the code developer will limit the specific kinds of information, formats, or values a user can use in the parameter.

Parameter Data Types

ArcGIS includes a large number of data types that a code developer can choose from. For a complete list the reader is directed to the ArcGIS document found in the **\ArcGIS\Desktop10.0\Documentation\ Geoprocessing_data_types.pdf** file and in the ArcGISH Help topic **Geoprocessing—Geoprocessing tool reference— Geoprocessing tools supplementary tool parameters**.

Parameter Properties

Parameter Properties can be modified to limit the values filled into the parameter to the proper data types, formats, values, etc. chosen by the tool's user. The Parameter Properties include the following:

a. *Type* – (**Required, Optional, Derived**). Indicates if this parameter is *required* for the tool to function, *Optional* where the user can set it or not, or *Derived*, which indicates if the parameter will be derived from another parameter and will NOT be shown in the tool's dialog box. For example, an intermediate feature class that is created as part of the process may be derived and will not be visible to the user interacting with the tool. Derived data types are always *Outputs*. (See Direction below).
b. *Direction* – (**Input or Output**). Inputs are datasets required to process the tool routine. Outputs are datasets that are used as intermediate datasets or as final datasets that will result when a tool routine completes executing (such as a feature class, table, or image).
c. *Multivalue* – (**Yes or No**). If a list of options for a parameter can be chosen, then set this parameter is set to Yes. An example of a multivalue may include choosing multiple values from a list (through a check box and a Value List) or adding multiple feature classes to a list. See the ArcGIS help document for more details (e.g. **Data Management Tools—General—Append** or **Analysis Tools—Overlay—Spatial Join**).
d. *Default* – If a code developer wants to have a default value show up in a tool parameter, then this property will be assigned a value. If the value is in a *Value List*, make sure consistent name syntax is used. For example, if a *Value List* contains the value 'FEET', then 'FEET' will be used in the default property and not 'Feet'. (This property is also good for file types such as *.dbf).
e. *Environment* – if the parameter references an environment setting, then this value is populated. For example, an Extent, Cell size, Pyramid Layers, etc. The screenshot below shows some of the options that are possible for the *Environment* property.

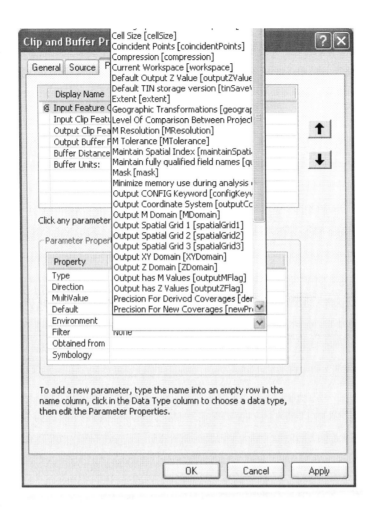

f. *Filter* – (None, Value List, Range, Feature Class, File, Field, Workspace). A filter is a condition that can be placed on the parameter to limit its options. For example, if a parameter *Type* is a feature class, then the *Filter* property can be set to *'Feature Class'* to only allow feature classes to be used for this parameter. This can help the end user choose appropriate values for parameters. After the *Filter Type* is chosen a space to add specific values or check boxes for specific types can be used. Click on the *'…'* (ellipse) when it appears in the *Filter* option. The following table summarizes the different *Filter Types* and possible values that can be used for each type.

Filter Type	Values
Value List	Typically strings or numbers For example, ('Feet', 'Meters', 'Yards', etc.) or (1,2,3,4,5, etc)
Range	Minimum and Maximum Ranges for numbers. For example, 0-10
Feature Class	Can be any supported feature type. Possible options that show up as check boxes include: Point, Multipoint, Polygon, Polyline, Annotation, Dimension, Sphere, Mutipatch. More than one type can be chosen.
File	A list of file suffixes (TIF, IMG, TXT, CSV, SHP, etc.). A list of values is typed in by the user using semicolons to separate suffix types and does not include the '.' (dot)). For example: `tif; img; txt; csv; shp`, etc.
Field	List of allowable field types such as Short, Long, Blob, Raster, GUID, etc. More than one value can be chosen.
Workspace	List of allowable workspace types, such as the file system (shapefiles, ArcInfo Coverages), local databases (file or personal geodatabase, or remote databases (ArcSDE geodatabases). More than one type can be chosen.

g. *Obtained from* – this property is used to set a dependency on another parameter. For example, a *Field* parameter may have the *Obtained from* property set to a *feature class* or *table* parameter to indicate that the fields come from the respective feature class or table. Fields represent the attribute table in a feature class or table.

h. *Symbology* – used only with output parameters and points to the location of a (`.lyr`) file and is used for displaying the output in ArcMap

Referring to the **Clip and Buffer Tool** Properties dialog box below, the Parameter Properties for the Buffer Units (string data type) is limited to a list of specific values (Filter property is set to a Value List) and the Default value is 'Feet'.

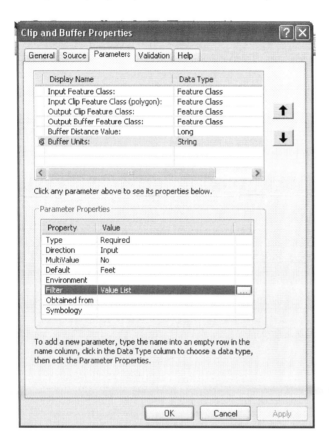

The Value List contains the following values:

   Feet, Meters, Kilometers, Yards, Miles

The following values are used to set up the parameters for the **Clip and Buffer** script tool.

Parameter Table for the Clip and Buffer Tool

Display Name (Parameter)	Data Type	Parameter Property	Value	Additional Settings
Input Feature Class:	Feature Class	Use defaults		
Input Clip Feature Class (polygon):	Feature Class	Filter	Feature Class	Check only Polygon from list
Output Clip Feature Class:	Feature Class	Use defaults except Direction	Direction = Output	
Output Buffer Feature Class:	Feature Class	Use defaults except Direction	Direction = Output	
Buffer Distance Value:	Long (Integer type)	Use defaults	(optional) set Default to a value (such as 100)	*This value might be useful, depending on the kind of buffer is required.
	String	Filter	Value List containing the values (Feet, Meters, Kilometers, Yards, Miles)	Click the Value List name or '...' to add the specific values to the list

The **Clip and Buffer** tool interface will look like the following after all of the properties have been set.

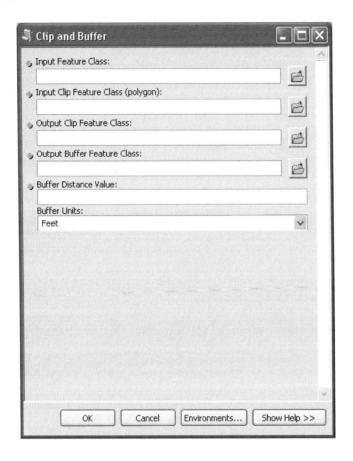

Customizing the Script Tool Interface

The discussion above illustrates the primary parts of the script tool interface that can be created and changed. The specific modifications to the **Clip and Buffer** tool can be found in Demo 10. Depending on the type of tool created, the programmer determines the kind of parameters and properties associated with each parameter.

Modifying the Python Script

To run a Python script from a tool interface, the Python script needs some minor modifications. Up to this point, the code developer has been writing and testing code by setting specific workspaces, data paths, feature class names, and other parameters so that the script can be run from the Python IDLE or command line. Since this script is ready to be deployed to an end user in ArcGIS, some changes need to be made to some of variables used in the Python script.

To accept values from an ArcGIS tool, the script must name variables differently instead of "hard coded" string variables. They must be changed to a `GetParameterAsText()` routine or Python system argument. The `GetParameterAsText()` routine is an ArcGIS construct that uses an index for each argument so that parameters can be passed from the tool interface to a variable in the script. For additional information, the reader can refer to the ArcGIS help document: **Geoprocessing—Creating tools—Creating script tools with Python scripts—Understanding script tool parameters**.

Adding Python Script Parameter Arguments

Python can accept script parameter arguments so that the parameters can be typed or passed from an application (in this case ArcGIS) to the script. The script parameter argument takes the form.

```
variable_name = arcpy.GetParameterAsText(<index>)
```

where `index` is the parameter number in the custom ArcGIS tool beginning with zero (0). The first parameter in the Tool user interface is set to `arcpy.GetParameterAsText(0)`. All of the tool parameters that are expected to be assigned to variables in the Python script will be ordered from top to bottom in the tools parameters. For **Clip and Buffer** tool example, the input feature class is the first parameter and thus will be the first script parameter argument in the Python script, `arcpy.GetParameterAsText(0)`.

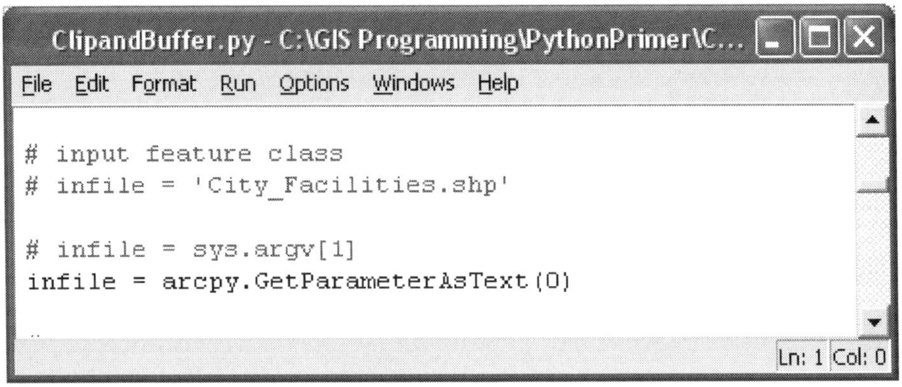

```
# input feature class
# infile = 'City_Facilities.shp'

# infile = sys.argv[1]
infile = arcpy.GetParameterAsText(0)
```

NOTE: The above figure shows an alternative method (commented out) for accepting parameters from a script tool (`sys.argv[1]`). The reader is encouraged to refer to the ArcGIS Help **Geoprocessing—Creating tools—Creating script tools with Python scripts—Understanding script tool parameters** for additional information. Examples of both methods can be found in the demo scripts the author's solutions. Either method is valid, provided the proper syntax is used.

This syntax will allow any valid feature class to be entered by the user which will then be assigned to the `infile` variable.

The next parameter in the **Clip and Buffer** tool is the input clip polygon feature class that will be used to clip the input feature class. Since this is the second parameter, it will be assigned to `arcpy.GetParameterAsText(1)` which is then assigned to the `clipfile` variable.

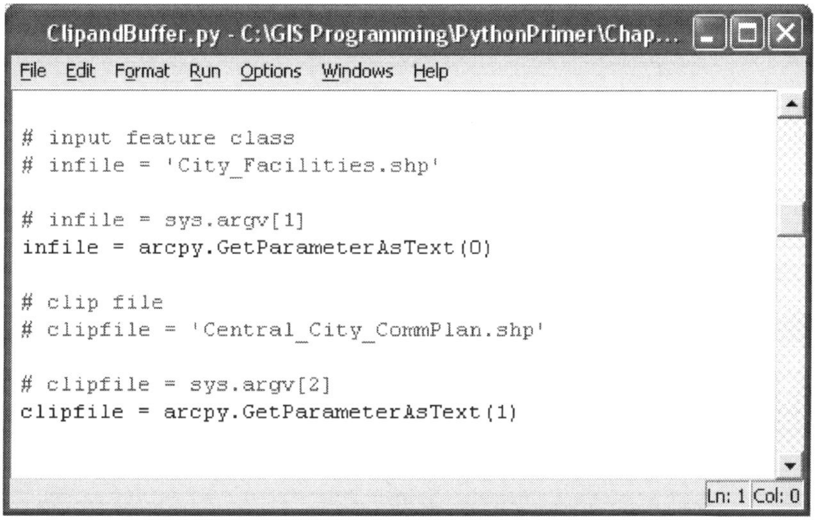

ClipandBuffer.py - C:\GIS Programming\PythonPrimer\Chap...
File Edit Format Run Options Windows Help
# input feature class
# infile = 'City_Facilities.shp'
# infile = sys.argv[1]
infile = arcpy.GetParameterAsText(0)
# clip file
# clipfile = 'Central_City_CommPlan.shp'
# clipfile = sys.argv[2]
clipfile = arcpy.GetParameterAsText(1)
Ln: 1 Col: 0

The rest of the parameters can be changed in a similar manner. The code developer should note the name of the variable in the script and which tool parameter it will accept values from. The code developer may choose to re-order the variable so they correspond to a similar order in the ArcGIS toolbox. The illustration below shows the rest of the variables and their respective script parameter arguments.

```
# output path
# Not needed since the full path to the output feature class are par
# the output feature class parameters.  Seee below and the tool inte

# outpath = 'C:\\GIS Programming\\PythonPrimer\\Chapter04\\MyData\\'


# output feature class
# outfile = outpath + 'City_Facilities_Clip.shp'

# outfile = sys.argv[3]
outfile = arcpy.GetParameterAsText(2)


# output buffer feature class
# outbuffer = outpath + 'City_Facilities_Buffer.shp'

# outbuffer = sys.argv[4]
outbuffer = arcpy.GetParameterAsText(3)

# buffer distance
# buff_val = 100

# buff_val = sys.argv[5]
buff_val = arcpy.GetParameterAsText(4)

# buffer units
# buff_units = 'Feet'

# buff_units = sys.argv[6]
buff_units= arcpy.GetParameterAsText(5)

buff_dist = str(buff_val) + ' ' + buff_units

try:
```

Notice that the `buff_dist` variable is not assigned to a parameter since it is made up of the buffer value (`buff_val`) and the units (`buff_units`) that are acquired from the user. The reader can also check the order of the parameters in the **Clip and Buffer** tool with the index value for each of the script parameter arguments noted above.

NOTE: The additional comments shown in the code above can be left in when the programmer is developing a script. These comments can be helpful to track previous attempts for troubleshooting and working on problems.

Adding ArcGIS Messages to the Python Script

Before completing this chapter about creating a custom tool to run a script, it will be helpful if some ArcGIS messages (a.k.a. the `print` statements that have been used up until now) can be reported back to the tool's progress dialog box so the tool user can see if progress is being made when the tool is executed.

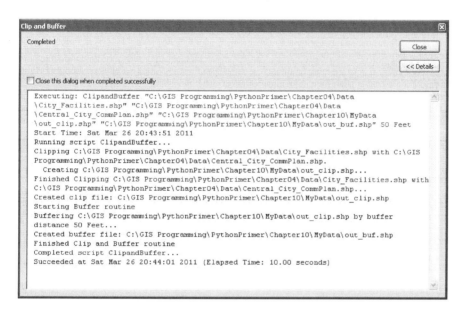

Up to this point, the scripts have used `print` statements to print messages only to the Python Shell.

ArcGIS messages can be easily added to a Python script by using the following syntax.

```
arcpy.AddMessage('<text string>')
```

For the **Clip and Buffer** script tool, the following messages were added to the script. See the script below.

```
buff_dist = str(buff_val) + ' ' + buff_units

try:

    # check to see if feature class already exists
    # if it does, delete it
    if arcpy.Exists(outfile):
        arcpy.Delete_management(outfile)

    print 'Starting Clip routine'
    arcpy.AddMessage('Clipping ' + infile + ' with ' + clipfile + '.\n Creating ' + o

    # Parameters using variables (Notice the indent because of the try statement)
    arcpy.Clip_analysis(infile, clipfile, outfile)

    print 'Finished Clip routine'
    arcpy.AddMessage('Finished Clipping ' + infile + ' with ' + clipfile + '...')
    arcpy.AddMessage('Created clip file: ' + outfile)


    if arcpy.Exists(outbuffer):
        arcpy.Delete_management(outbuffer)

    print 'Starting Buffer routine'
    arcpy.AddMessage('Starting Buffer routine')
    arcpy.AddMessage('Buffering ' + outfile + ' by buffer distance ' + buff_dist + '..

    arcpy.Buffer_analysis(outfile, outbuffer, buff_dist)

    print 'Created buffer file'
    arcpy.AddMessage('Created buffer file: ' + outbuffer)

    arcpy.AddMessage('Finished Clip and Buffer routine')

except:
    print arcpy.GetMessages(2)
    tb = sys.exc_info()[2]
```

The above script shows both the `print` and the `AddMessage` statements. Only the `AddMessage` statements are processed and reported back to the process dialog box when the ArcGIS user uses the tool graphical user interface. The `print` statements are provided as the programmer initially developed the code. These can be left in and used if the programmer needs to further develop the scripts before "re-deploying" the custom tool to ArcMap (or ArcCatalog). They do not need to be commented out.

The `AddMesssage` function prints text statements to the Progress dialog box as processes are completed. The text statements can also include variable names and values reported from other routines such as the number of selected features or a count of objects.

In addition to the `AddMessage`, two other kinds of messages can be reported back to the Progress dialog box:

1. `AddWarning`
2. `AddError`

`AddWarning` represents a message associated with a potential problem with the process, but does not necessarily indicate that the process will fail. Examples of warnings may include a select layer routine that does not include any selected features or a feature class that does not include a defined spatial reference. `AddError` is typically associated with a process that will cause a script or tool to fail. Examples of errors might include a looping structure that loops beyond a valid set of values or the feature class cannot be found or does not include the correct type of field values (e.g text, numbers, etc).

`AddWarning` and `AddError` messages can be added and customized by the programmer to provide meaning feedback to the end user of the tool. The **Exception.py** script referenced in the book uses the `AddError` message statement to report back a custom message when an ArcGIS error is encountered.

Executing the Script Tool

Once all of the parameters have been set and the Python script modified to include the `GetParameterAsText()` and the `AddMessage` routines, the script tool created in the ArcToolbox can be executed. The user should be able to fill in the parameters and click OK. If the script was written correctly with the correct syntax for the parameters and the messages, a progress dialog box should appear that reports the input values as well as any custom message added to the script. The following dialog box shows the parameters and messages that are used and written for the **Clip and Buffer** tool.

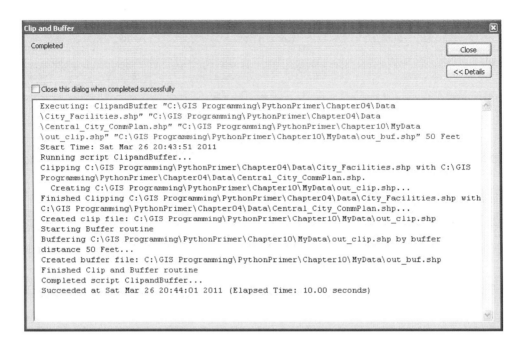

Any errors encountered will also show up in the progress dialog box because of the `traceback` module and message statements found in the except block of the script. The reader should note the use of the `AddMessage` and `AddError` routines.

ClipandBuffer.py - C:/GIS Programming/PythonPrimer/Chapter10/ClipandBuffer.py

File Edit Format Run Options Windows Help

```
    print 'Starting Buffer routine'
    arcpy.AddMessage('Starting Buffer routine')
    arcpy.AddMessage('Buffering ' + outfile + ' by buffer distance ' + buff_dist + '..

    arcpy.Buffer_analysis(outfile, outbuffer, buff_dist)

    print 'Created buffer file'
    arcpy.AddMessage('Created buffer file: ' + outbuffer)

    arcpy.AddMessage('Finished Clip and Buffer routine')



except:
    print arcpy.GetMessages(2)
    tb = sys.exc_info()[2]
    tbinfo = traceback.format_tb(tb)[0]
    pymsg = "PYTHON ERRORS:\nTraceback Info:\n" + tbinfo + "\nError Info:\n     " +
    msgs = "arcpy ERRORS:\n" + arcpy.GetMessages(2) + "\n"

    arcpy.AddError(msgs)
    arcpy.AddError(pymsg)

    print msgs
    print pymsg

    arcpy.AddMessage(arcpy.GetMessages(1))
    print arcpy.GetMessages(1)
```

Ln: 25 Col: 27

The following figure shows some error messages that were produced as a result problems encountered when the **Clip and Buffer** routine was executed. Note that the `except` block was implemented and specific error messages were then printed back to the progress dialog box.

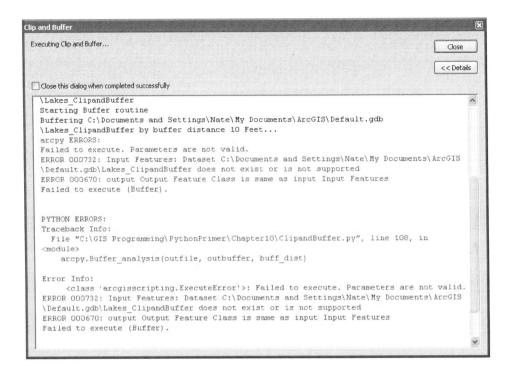

Writing Tool Documentation

Tool documentation is an important part of deploying a custom tool to an ArcGIS user. The documentation can provide some useful information and context that the end user can refer to for additional assistance when using the tool. Tool documentation can include brief descriptions of tool functionality, definitions and syntax of parameters, script examples and syntax, and illustrations. This section provides a brief introduction to develop tool help for a custom tool. The reader is encouraged to refer to the ArcGIS help document for more information **Geoprocessing—Creating tools—Documenting tools**.

NOTE: It is highly recommended that any modifications to the custom tool help are performed within ArcCatalog or the Catalog Window (in ArcMap). Either of these locations will provide the ability to edit all aspects of the custom tool help documentation (such as summary descriptions, descriptions of the individual parameters, as well as adding image illustrations, etc.). **DO NOT access the Item Description through the ArcMap Toolbox. Not all features are available and ArcMap may lock up.** *Any tools that have been developed with older versions of ArcGIS used different methods of editing tool help documentation. The existing documentation should appear, but some of the documentation may need to be rewritten or modified to work with ArcGIS 10.*

The code developer can create custom help documentation by locating the script tool in ArcCatalog and then click the Description Tab or by right-clicking on the script tool in the Catalog Window and selecting **Item Description**. Both methods are shown below.

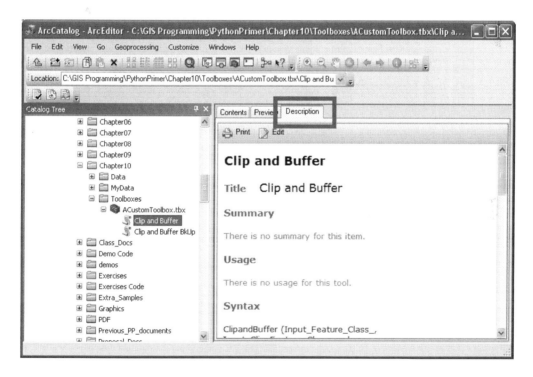

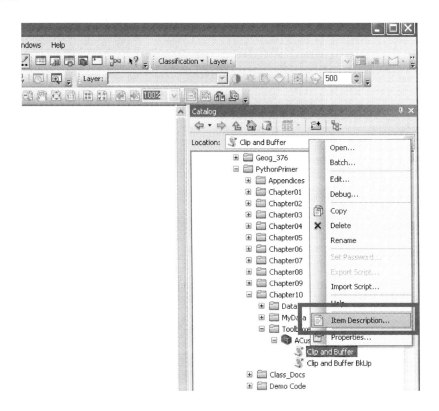

To edit the help document click on the **Edit** button at the top of the **Description** or **Item Description** tab. The following view appears showing a variety of areas that can be edited.

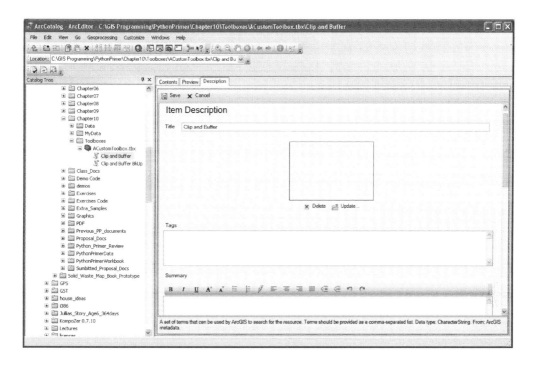

The title is already populated from the tool name provided when the tool was originally created. The help document has the following sections.

a. *Title* – title of the tool
b. *Tags* – comma separated keyword names that can be used by search tools
c. *Summary* – brief summary of the tool functionality
d. *Usage* – a brief overview of how the tool can be used. Refer to other tool help for ideas.
e. *Syntax* – specific help documentation for each parameter in the tool.
f. *Code Samples* – excerpts of code samples; similar to those found in other ArcGIS geoprocessing tool help documentation
g. *Credits* – Author of the script and/or tool
h. *Use Limitations* – brief description of any limitations of use for the script or tool

The documentation is easy to update by simply typing in descriptive information into each section.

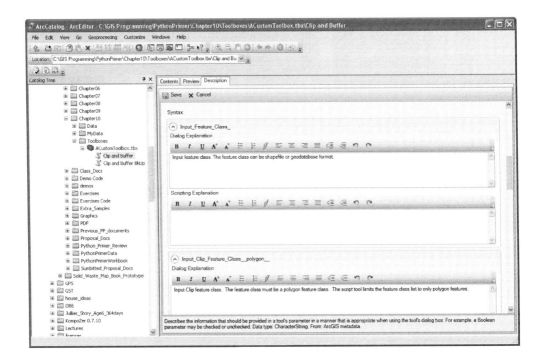

The tool *label names* (e.g. *Input Feature Class:* from the **Clip and Buffer** tool example) will show underscores "_" for any special characters used in the parameter label name. For example colons and parentheses will show up in the help as underscores. The label names cannot be changed in the help. See the Dlg.zip file that contains the custom help document created for the **Clip and Buffer** tool.

When the documentation is complete, it can be saved by clicking on the **Save** button. A preview can be reviewed in ArcCatalog or the Catalog Window.

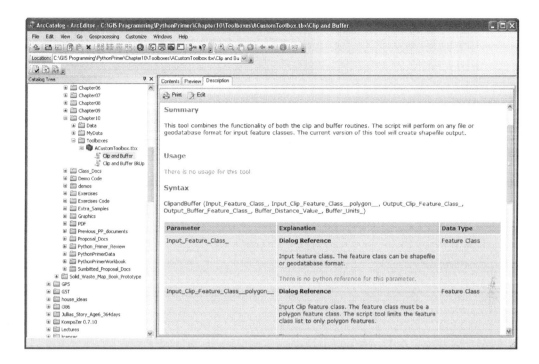

Notice that the documentation provided shows up in the respective portion of the help document.

The custom script tool can be opened where the **Tool Help** will show the updated help information.

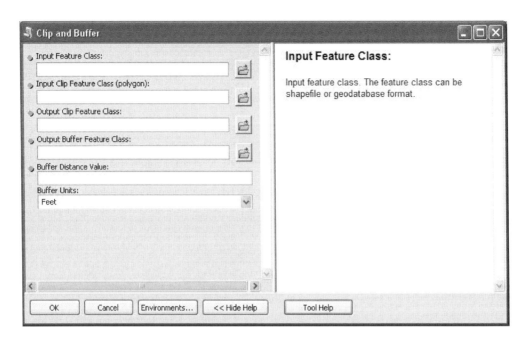

Click on the **Tool Help** button to show the complete help document for the script tool.

Note the location of the help document in the browser address field. The documentation is an HTML document that is written to the local systems ArcGIS toolbox under the specific user's Documents and Settings folder.

file:///C:/Documents and Settings/<user>/Application Data/Esri/Desktop10.0/ArcToolbox/Dlg/MdToolHelp.htm

The code developer can change this location if the tool and documentation need to be placed in a centralized location.

Summary

This chapter has focused on how to execute a script through a custom toolbox in ArcMap. Having a user interface where specific parameters can be added provides some additional flexibility with executing the script whereas, up to this point, the Python script has had to be executed through Python IDLE or similar Python editor. In addition, a tool separates a user from the code where the user may not have the experience or knowledge to work directly with the script.

The key elements to creating a functioning tool interface is to know which parameters are required to execute the script and which order the parameters need to appear in the tool. In addition, the code developer needs to set the correct data type and determine if a list of values needs to appear to a user, if default values are needed, if a parameter is an input or output, and if there are any dependencies of one parameter on another (e.g. a field list must come from a feature class or table). Each of these can be challenging above the challenges of writing the actual Python script. It is highly recommended that a Python script is thoroughly tested, error handling is added, and various conditional statements are added to check for any specific issues that might occur (e.g. does a spatial reference exist, is a feature class a polygon type, etc.) before deploying a script tool. Even after these kinds of quality checks are implemented, others may be discovered when designing and testing a tool graphical user interface (GUI) with the Python script.

Adding messages, warnings, and errors can also be added to provide some additional reporting and feedback to the tool user so that they can review parameters, datasets, and provide some basic trouble shooting before communicating with the code developer. The code developer can create specific help documentation for the custom script tool which can be used by end users of the tool.

Demo 10: Create a Custom Script Tool Interface for the Clip and Buffer Tool

This demonstration uses the **ClipandBuffer.py** scrip that has already been developed. The steps below can be followed to develop the graphical user interface (GUI) and parameter properties for the **Clip and Buffer** tool described in the chapter. Refer to the **Parameter Table** found in Chapter 10 under the **Parameter Properties section** to construct and define the parameters and properties for the tool interface. The reader can refer to the chapter to study and practice developing a custom script tool and the associated changes to the script as well as develop the help documentation. The script, toolbox (**ACustomToolbox.tbx**), data, and help document (**Dlg.zip**) can be found under **\PythonPrimer\Chapter10**.

The demo script (**ClipandBuffer.py**) contains the `GetParameterAsText()` and `AddMessage()` routines to pass values from the tool interface to the Python script and to write messages back to the progress dialgog box. Consult the script for more details.

The **Dlg.zip** file contains the help documentation created by the author for this tool. The reader is encouraged to generate their own tool help by using the methods described in this chapter to gain additional experience.

Create the Custom Toolbox

1. Before developing the tool interface, make sure to create a custom toolbox. Use the directions in the chapter to create the toolbox and then right-click to **Add--Script** to bring up the default dialog box for the tool properties.

2. Assign the tool a name **Clip and Buffer**. Make sure not to use spaces in the *Name* property. The *Label* can contain spaces. Provide a brief description if desired.

Assign the Python Script to the Tool

Click on the Source tab and find the **ClipandBuffer.py** script.

Customizing the Script Tool Interface Parameters

To create the script tool interface, the Display Name, Data Type, and Parameter Properties must be set. This section will illustrate the steps to set up the **Clip and Buffer** script tool interface parameters. The values in the above table will be used to create the interface.

1. Bring up the script tool Properties and click on the Parameters Tab. If this is the first time to create the toolbox and associate the script, the Next button is used to access the Parameters tab. If the code developer is modifying a script tool that does is already associated with a custom toolbox, but does not contain any parameters, right click on the script tool and choose Properties and then click on the Parameters Tab.

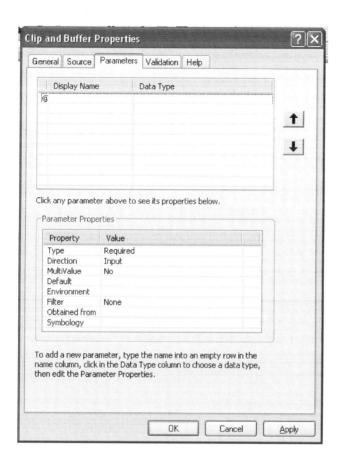

2. In the **Display Name** field enter the first Display Name from the information found in the table in Chapter 10 (e.g. **Input Feature Class**). The "@" symbol indicates which Parameter is being created or modified.

3. Click in the **Data Type** and choose *Feature Class* from the list. Keep the default values for all of the other Parameter Properties for this parameter.

4. When the code developer enters the **Clip Feature Class** parameter, the *Filter* Parameter Property is used. Enter the Display name and Data type as describe above. Next, in the *Filter* property select *Feature Class* from the **Value** column.

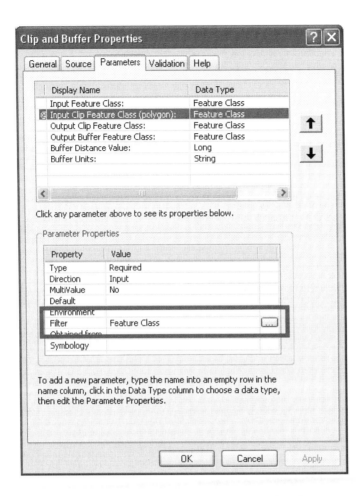

When *Feature Class* is selected, the following dialog box appears. Check only the *Polygon* check box since clip feature must be a polygon feature type.

NOTE: If the user needs to modify this value again, the Feature Class value can be clicked in the Value column or the "..." can be clicked to bring up this dialog box.

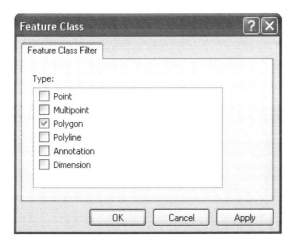

5. Other parameters can be added using the **Parameter Table** in Chapter 10. For both the **Output Clip Feature Class** and **Output Buffer Feature Class** change the **Direction** to *Output*, since these feature classes will be saved as outputs from the tool.

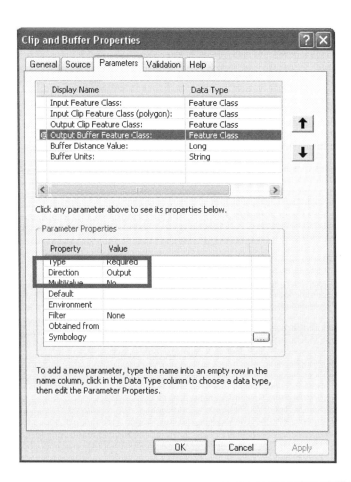

The **Buffer Distance Value** should be entered as a **long**, since the tool should only accept number values for this parameter. This will prevent the tool from accepting text characters. The **Buffer Units** parameter will be entered as a **String** data type and use a **Value List** for the **Filter**. When the code developer clicks on **Filter—Value List** a dialog box pops up so that the specific values for the list can be entered. The specific values for the unit types entered are:

    Feet
    Meters
    Kilometers
    Yards
    Miles

The resulting value list will look like this.

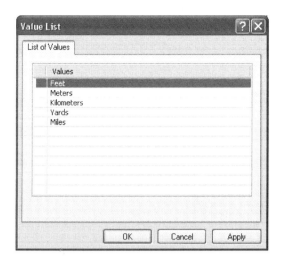

After clicking OK, the code developer can set the **Default** Parameter Property to **'Feet'**. This will provide **'Feet'** as the default value in the tool interface.

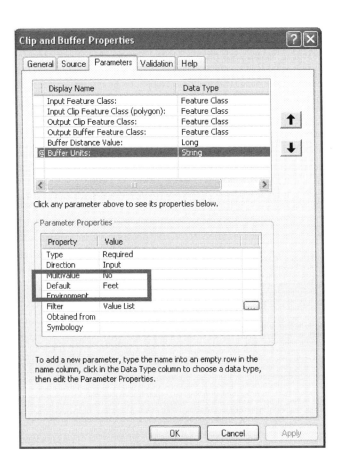

6. Now that all of the parameters have been set up, the code developer can click OK. The **Clip and Buffer** script tool can be opened by double clicking on the tool to see the new tool interface.

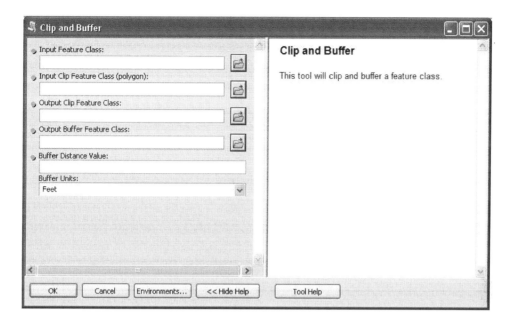

Notice the parameter's headings and see that all of the parameters are required (as a result of the "green dot" appearing next to each parameter), and that the **Buffer Units** parameter already has the default value loaded. If the dropdown list was clicked on, the user would see the other unit values that could be chosen for this parameter.

The reader can review the tool help developed for the **Clip and Buffer** tool and can experiment with editing the documentation. Refer to Chapter 10 for an overview of how to develop and modify the custom help documentation.

Exercise 10: Create a Custom Script Tool Interface for Your Own Script

This exercise uses the concepts and methods above to generate a tool interface for a script that the reader has generated. It is suggested that the Chapter 7 script for **Batch Clipping Images** be used as a starting point for this exercise. (Any existing script can be used that the reader chooses). The reader will need to study the script developed in Chapter 7 and determine which parameters are required to execute the script through the script tool. In addition, the reader can develop the help document for this tool. At the completion of this exercise, the reader should have a custom tool interface that successfully mimics the same functionality as the script that is run through the Python IDLE editor.

Chapter 10: Questions

1. What key elements are required to successfully create and implement a custom script tool?

2. True/False Is a set of parameters absolutely required to run a Python script? If False, describe how this is possible and what tool interface elements are not present in the tool interface.

3. What elements are required to create a specific tool parameter?

4. Briefly describe what the Parameter Properties are? How are they used to modify or limit a specific tool parameter?

5. Briefly describe the following and how they are used with a tool interface.

    a. `GetParameterAsText()`
    b. `AddMessage()`
    c. `AddWarning()`
    d. `AddError()`

6. What are the two locations in ArcGIS where a code developer can create and edit help documentation?

7. When the tool help is updated, how does a tool user access or see the tool help?

Chapter 11   Automating Geoprocessing Scripts

Overview

Throughout *A Python Primer for ArcGIS* the scripts have primarily been executed through the Python IDLE window or other Python editor. Chapter 10 illustrated another method for implemented a Python script to perform geoprocessing tasks by using a custom ArcGIS toolbox. One of the primary reasons for writing Python scripts is to automate repetitive geoprocessing tasks that normally take considerable manual interaction with ArcMap or ArcCatalog. Throughout the book a number of common geoprocessing tasks have been illustrated that shows how to set up the tasks in a logical order so that they can be "automatically" executed. Up to this point one requirement to execute Python scripts has been to:
1. A user must click Run in Python IDLE
2. Fill in parameters to run a custom geoprocessing script tool

An alternative method not discussed is using the "command line" to execute a Python script. This involves typing in the syntax to execute a Python script as well as the specific parameters the script requires at a Command Prompt window. (See the figures throughout this chapter that show the Command Prompt window). This chapter illustrates this method, since it is required to automatically run (batch process) a Python script.

A "batch file" will be used by the Windows Scheduler to run a Python script at a specified time and/or frequency where the code developer or user do not have to physically run the script. The batch file contains all of the information to automatically run a geoprocessing script.

The Batch File

Automated processing of programming routines is a common practice among programmers and is often performed by the use of a batch file. Essentially, a batch file is a collection of one or more executable statement to run any number of processes without the user interacting with a specific program or routine. These could be data management activities such as copying data to different directories or folders, run executable commands with a variety of parameters, or to run Python scripts that perform geoprocessing tasks that may also include parameters that used by the script. Because Python files are "executable," they function in a similar manner as other executable (.exe) files. Essentially, the batch file contains all of the required parameters to run the Python script.

Running a Python Script at a Command Prompt

As mentioned before, Python scripts can be executed through the Command Prompt. The Command Prompt is a window that contains a drive letter and any of the folders (directories) it contains. Before Windows environments were main stream to computers, the Command Prompt was the primary interface to interact with the computer. The Command Prompt persists so that computer users and programmers can execute (or run) commands and programs in an alternative manner to clicking an icon, or navigating through a file management window (such as Windows Explorer) to locate a program and then execute it. Usually, the Command Prompt can be found under **Program Files—Accessories—Command Prompt** or the user can save a short cut to this window and place it on the desktop or in a launch bar at the bottom of the computer screen for easy access. The following figure represents the Command Prompt on a Windows operating system.

```
Command Prompt

Microsoft Windows XP [Version 5.1.2600]
(C) Copyright 1985-2001 Microsoft Corp.

C:\Documents and Settings\Nate>cd c:\gis programming\pythonprimer\chapter11

C:\GIS Programming\PythonPrimer\Chapter11>dir
 Volume in drive C has no label.
 Volume Serial Number is 8C51-82EE

 Directory of C:\GIS Programming\PythonPrimer\Chapter11

05/22/2011  12:28 PM    <DIR>          .
05/22/2011  12:28 PM    <DIR>          ..
10/24/2011  11:42 AM             2,191 CommandLine_Clip.py
05/22/2011  12:33 PM               178 CommandLine_Clip_BatchFile.bat
05/22/2011  01:28 PM    <DIR>          Data
05/22/2011  01:28 PM    <DIR>          MyData
03/28/2011  09:33 AM                80 PythonIsGreat.py
               3 File(s)          2,449 bytes
               4 Dir(s)  11,066,281,984 bytes free

C:\GIS Programming\PythonPrimer\Chapter11>
```

The Command Prompt shows the drive letter at the top of the screen and a set of folders (directories and subdirectories); in this case **C:\PythonPrimer\Chapter11**. The folder path indicates that the command prompt is at the *C drive* and in the directory *PythonPrimer* and the subdirectory *Chapter11*. The user can change locations in the Command Prompt by using common DOS commands (such as *CD* for *change directory*). If the user types in *DIR* (list the contents of the directory) a list of subdirectories and files can be seen. The reader is recommended to search the Internet for DOS commands to find out more information; however, for the purpose of creating batch files for automatically running Python scripts, all of the DOS commands that a code developer will typically use can be found in this chapter.

For the purposes of executing Python scripts, the user can change directories to the location of a Python script and type the following to execute a Python script.

```
<drive><path where python file exists>python <name of python script>
```

For example, in the `pythonisgreat.py` script that can be found in the **\PythonPrimer\Chapter11** folder, the following can be written at the Command Prompt. If the reader copied the files to his/her local file system in the "C drive", the following can be performed to run the script.

1. Bring up the Command Prompt (**Program Files—Accessories—Command Prompt**)
2. Type the command `C:\PythonPrimer\Chapter11 <enter>`
3. Type the command `DIR <enter>`. Make sure the **PythonIsGreat.py** file is listed in the directory (folder)
4. At this prompt (i.e. the location shown in the Command Prompt window) type the following: `python pythonisgreat.py <enter>`

In the figure below, the directory (folder location) and the command `python` executes the `pythonisgreat.py` file.

C:\PythonPrimer\Chapter11>`python pythonisgreat.py`

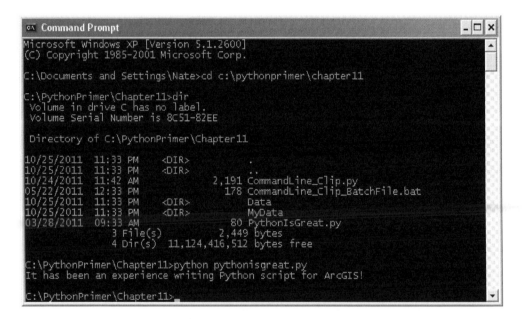

When the "Enter" key is tapped after typing in `python pythonisgreat.py`, the Python file is executed and prints the statement above. The word *python* actually runs the `python.exe` file which then executes the Python script file, `pythonisgreat.py`. Since the path to the *python.exe* file is in a user's computer path (i.e. the PATH environmental variable found under **My Computer—System Properties—Advanced Tab—Environment Variables**), the `python.exe` executable file can be run from any location simply by typing the word `python`. If only the word `python` is typed at the

Command Prompt, the Python Shell prompt appears. The user could begin typing in Python code and run individual Python commands.

If a script requires parameters, the command line syntax would show the individual parameters separated by spaces. In the example the `commandline_clip.py` script contains five parameters:

1. *Workspace* – path to the folder containing data (`'c:\\pythonprimer\\chapter11\\data'`)
2. *Input file* – `city_facilities.shp`
3. *Clip file* – `central_city_commplan.shp`
4. *Output workspace* – path to a folder for output (`'c:\\pythonprimer\\chapter11\\MyData'`)
5. *Output file* – `out_file.shp`

Running the script through the Command Prompt, the following general syntax is used:

```
<command prompt> python <name of script> <parameter 1> <parameter 2> <...>
```

Notice each parameter is separated by spaces. For the `commandline_clip.py` script, the command line syntax looks like the following figure.

```
10/26/2011  12:07 AM    <DIR>          .
10/26/2011  12:07 AM    <DIR>          ..
10/26/2011  12:05 AM             2,753 CommandLine_Clip.py
05/22/2011  12:33 PM               178 CommandLine_Clip_BatchFile.bat
10/26/2011  12:05 AM    <DIR>          Data
10/26/2011  12:05 AM    <DIR>          MyData
10/26/2011  12:07 AM                 0 python
03/28/2011  09:33 AM                80 PythonIsGreat.py
               4 File(s)          3,011 bytes
               4 Dir(s)  11,065,774,080 bytes free

C:\PythonPrimer\Chapter11>python commandline_clip.py c:\pythonprimer\chapter11\d
ata city_facilities.shp central_city_commplan.shp c:\pythonprimer\chapter11\myda
ta out_clip.shp

Running Clip Routine...
Input workspace is: c:\pythonprimer\chapter11\Data
Input file is: city_facilities.shp
Clip file is: central_city_commplan.shp
Output workspace is: c:\pythonprimer\chapter11\mydata
Output file is: c:\pythonprimer\chapter11\mydata\out_clip.shp
Completed Clip

C:\PythonPrimer\Chapter11>
```

The command line syntax begins with `python` and then the name of the script `commandline_clip.py` and then each parameter separated by a space. The list of parameters will span multiple lines. All of the commands are typed at the command prompt at one time.

NOTE: If spaces exist in the directory path, then the path string will need to include double quotes so that Python interprets the string as a single parameter. For example, the following path contains spaces which are bounded by double quotes.

`"c:\gis data\pythonprimer\chapter11"`

The code developer should note that if the Python script requires parameters that "get" values (e.g. the lines of code that use the `arcpy.GetParameterAsText(n)` routine, where *n* is the argument number), that these parameters are included in the command line arguments. The table below summarizes the script parameters and the values typed at the command prompt.

Python Script Variable	Python Script Parameter arcpy.GetParameterAsText(n)	Value typed at Command Prompt
arcpy.env.workspace	0	C:\pythonprimer\chapter11\data
infile	1	city_facilities.shp
clipfile	2	central_city_commplan.shp
output_ws	3	c:\pythonprimer\chapter11\mydata
outfile	4	out_clip.shp

Creating a Python Batch File

The batch file is simply a text file that contains the command line syntax for running a Python script similar to the above figure. The file name must end with the word '.BAT' which the Windows operating system recognizes as an executable batch file.

NOTE: Any file on a Windows operating system with the .BAT extension is executable.

The batch (.BAT) file can be created by using a text editor (such as Notepad) and writing the complete command line syntax and saving the file with the .BAT extension. It is recommended to use a generic text editor versus Word or a specific document writer, since these programs may add special characters that are invisible to the user (such as line returns, paragraph breaks, tabs, etc.).

NOTE: The Python command file and script parameters are written on a single line with the parameters separated by a space.

The following screen shot shows the same command line syntax for the above Python script that is written In a notepad. The batch file itself has been named **CommandLine_Clip_BatchFile.bat**. This file can be found in the Chapter11 folder. Use a generic text editor to open the file and see the specific syntax.

The **CommandLine_Clip_BatchFile.bat** file can be "double-clicked" in the Windows Explorer to run the Python script. When the script is executed within the Windows Explorer, a command line window appears and shows the syntax of the program. Alternatively, the user can open the Command Prompt window and change directories (*CD*) to the location of the batch file (.BAT) and then type the name of the .BAT file at the command prompt to run the batch file.

If any print messages exist within the script, they will print in the command line window.

NOTE: The command line window will close automatically when the script successfully executes. If the user wants to keep track of the print messages as the script executes, the print statements can be written to a log file. See Chapter 8 for creating and using a log file.

The following figure shows the print statements that print to the command line window before it automatically disappears.

Scheduling the Batch File to Automatically Run the Geoprocessing Script

The batch file developed above must be clicked on in a Windows Explorer or typed in a command line prompt for the script to execute. To have the script run automatically on a Windows machine, it must be scheduled using the Windows Scheduler. The Windows Scheduler can be found in the **Control Panel— Scheduled Tasks**.

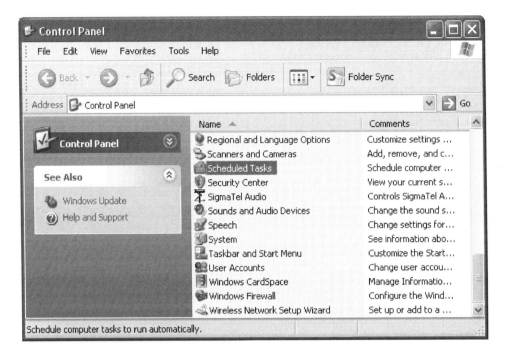

A task can be added and scheduled by clicking on **Schedule Tasks—Add Schedule Task**. See the figures above and below.

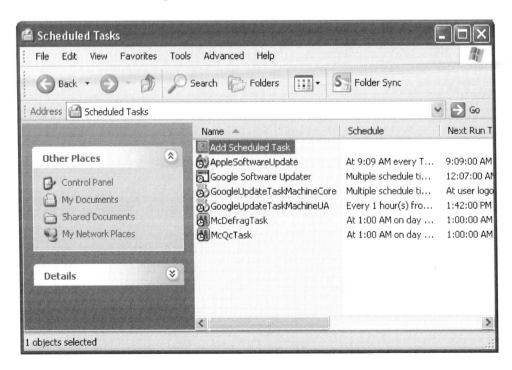

Use the **Scheduled Task Wizard** to schedule the specific task. When the wizard prompts for the specific task, click the Browse button and locate the batch file (.BAT) that contains the Python commands, script, and required parameters.

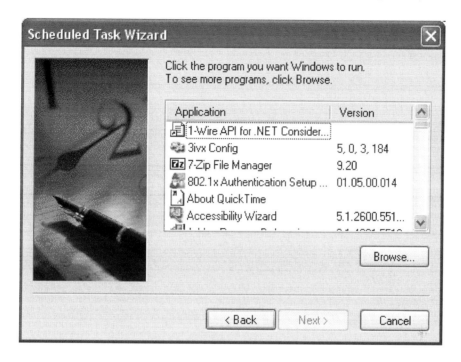

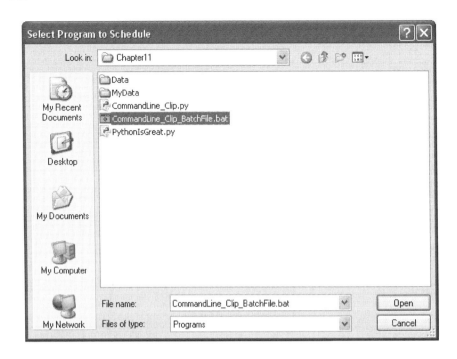

Change the name of the task if desired and then select the frequency for running the script. Some organizations use daily or weekly frequencies to automatically run scripts. See the figure below.

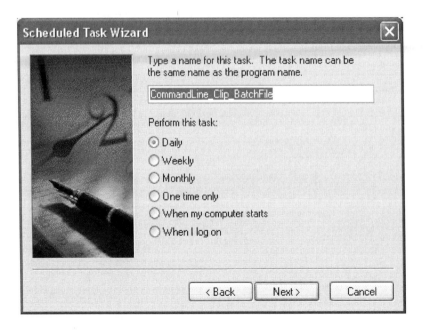

Select the specific time for the script to run. From an organizational point of view, the specific time may be dependent on how long the script takes as well as other schedule processes that may depend on the data being created or updated by a separate geoprocessing script.

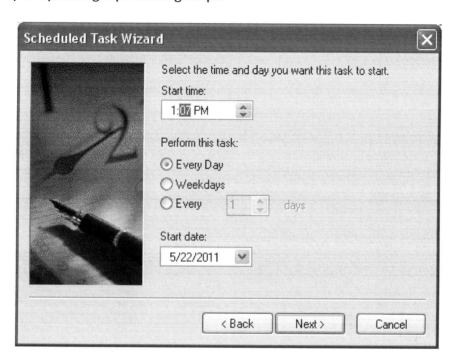

To properly schedule a task, a proper user name and password must be set. Normally, scheduled tasks are run on organizational servers that require proper login credentials. On a personal system the user can set up his/her respective set of credentials. The computer must have a password for the script to properly run or the scheduler set to run if the user is logged in. Not putting in a password will result in the scheduled task not being able to run and an Access Denied error will show when the user attempts to schedule the task without a password.

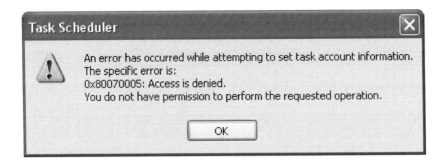

As an alternative, if an Access Denied message appears, the user can go into the Schedule Tasks Properties under the Task Tab and check the "Run only if logged in" button. This will alleviate the need to use a specific password. Also note that some settings may need to be changed if running the schedule task on a laptop (e.g. "Don't start the script if the computer is running on batteries", etc). For organizational systems it is recommended to use secure login credentials and proper backup systems are in place.

In addition, when testing script automation processes it is good practice for an organization to have a "test environment" and a "production environment." Test environments are used to test processes and methods and troubleshoot any pending issues with data, scheduling, and the processing of data. Only when a script and process has been thoroughly tested can it then be placed and scheduled on a "production environment," one that includes a server used to perform real business functions and real (live) data, and includes the proper information security and backup system or server redundancy.

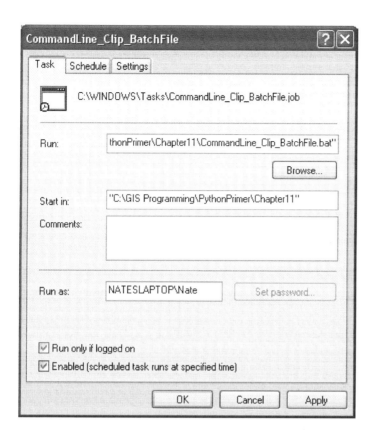

The new task will show up in the Scheduled Tasks window. The time and frequency will show and when the task was last run as well as when the task will subsequently run. See the figure below.

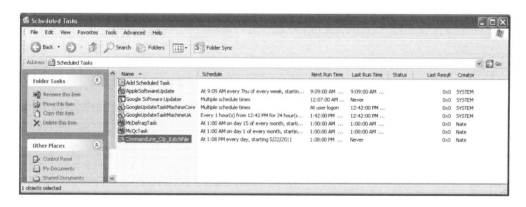

When batch file is actually running the following appears in the Schedule Tasks window. The batch file shows that the status is "Running." See the figure below. If the task runs successfully the "Last Result" column will show "0x0". If the schedule task fails, the "Last Result" column will show "0x1." The user should also review any log files that collect Python print statements to assist any trouble shooting efforts to remedy any scripting or potential network problems. Other system logs and computer administration may need to be contacted if required. If the scheduled task successfully executes, the Python script was automatically run. The scheduled task will continue to function properly unless changes to the process are required or problems with the machine or data server are incurred. If changes are made to the script and/or data changes, these should be tested on the test environment before deploying to the production environment.

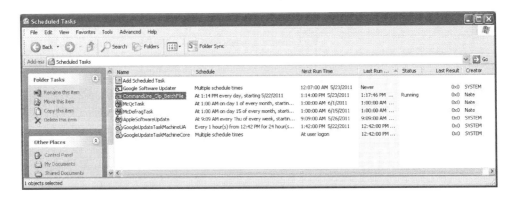

Summary

A Python script can be automatically run by using a Windows batch (.BAT) file and scheduling it in the Window Scheduler (**Scheduled Tasks**). Using these methods, geoprocessing Python scripts can run after hours or during off-peak hours without the GIS staff or program developer to physically interact with them. Being able to automatically executing ArcGIS geoprocesses can free an analyst's time to work on analytical, data management, compilation, or other mapping tasks. Automated geoprocessing tasks are often implemented to update relational databases, or file and publication servers that serve out data, maps, web services, and other geosptial and tabular information.

Demo 11: Create and Schedule a Batch File to Auto-run a Python Script

The following instructions will demonstrate creating, scheduling, and running a batch file for the script mentioned above on a Windows XP 32-bit machine. Windows 7 (32-bit or 64-bit) machines may have a slightly different interface for scheduling tasks. The script and data are located in **\PythonPrimer\Chapter11**.

1. Review the existing CommandLine_Clip.py script. Make sure to change the paths for the output and log folders to point to the Chapter 11 folders on the local system. Save the Python script.

2. Run the script from Python IDLE to make sure it runs correctly. Close all Python IDLE and Python Shell windows.

3. Open a text editor (such as Notepad) and add the following lines. Save the file as CommandLine_Clip_BatchFile.bat. The batch file includes the following parameters. Refer to the table in Chapter 11 for the order and values of the individual parameters shown below.

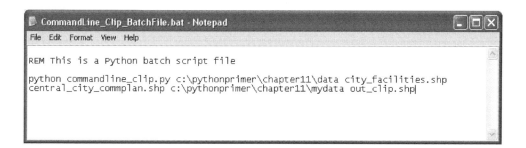

These parameters are specifically used in the Python script. Refer to the "GetParameterAsText" values in the script.

NOTE: The Python line shown above does not include any line returns and does include a space between each parameter. Make sure to enter the proper path to the shapefile shown unless the reader placed the PythonPrimer folder in a different location.

4. Test the batch file by opening Windows Explorer. Find and double click the `.BAT` file created above. Make sure the script executes successfully. If errors occur, check the syntax of the `.BAT` file to make sure it is correct. A successfully executed script will result in the Command Line window appearing and then disappearing.

5. Check the **MyData** folder to see that the **out_clip.shp** file was created with a current date and time stamp as well as a **log file** with the current date.

6. If the above is successful, open the **Scheduled Tasks** program from **Start—Control Panel—Scheduled Tasks**.

7. Click **Add Scheduled Task**. Follow the wizard to set up a scheduled task for the batch file created above.

8. Check at a later time to see if the batch process ran successfully. This can be checked by reviewing the date and time stamp of the files in Windows Explorer or ArcCatalog as well as checking the log file.

Chapter 11: Questions

1. What are 3 different ways to run a Python script?

2. What is a primary benefit of scheduling a Python script?

3. To schedule a script for automatic processing, what kind of file is used in the scheduler?

4. What kind of information must this file include?

5. If the Task Scheduler is used on a server, what information is likely needed to successfully schedule the script?

Appendices

Appendix 5.1   Python, Cursors, and Open Source Databases

For those interested in Python cursor methods for accessing databases (such as SQLite, http://sqlite.org/) can see the Python documentation (http://docs.python.org/release/2.6.5/library/sqlite3.html). SQLite3 is not part of Python, but a separate database application for performing database tasks. A SQLite3 set of Python libraries exist to perform a variety of database tasks, including the use of cursors. The reader can check out these references for more information on open source databases and GIS.

NOTE: These methods will likely not work on GIS type databases. If using databases such as Oracle or SQL Server, the organization may have already invested in ArcSDE, the middleware to perform GIS tasks on enterprise grade databases such as Oracle or SQL Server.

References for Open Source databases and GIS database functions, see the following:

>   PostgreSQL - http://www.postgresql.org/ - open source relational database
>   PostGIS - http://postgis.refractions.net/ - GIS components for PostgreSQL

Also, the reader can refer to these sites for MySQL, another open source relational database.

>   MySQL - http://www.mysql.com/
>   MySQl Spatial Extensions - http://dev.mysql.com/doc/refman/5.0/en/spatial-extensions.html

Appendix 9.1   Summary of the mapping Module Properties, Methods, and Functions

The following summarizes some map element properties, methods, and functions that are commonly used with the `mapping` module and Python. For a comprehensive list and description see the ArcGIS Help documents under **Geoprocessing—The ArcPy site package—Mapping module**. Searching ArcGIS Help for specific terms may be required.

Map Document

Properties

`<map_variable>.<property> = <value>`

e.g. `mxd.author = 'N. Jennings'`

```
author
relativePaths (True/False)
title
summary
filePath
description
```

Methods

`<map_variable>.<method>`

e.g. `map.Save(), map.SaveACopy(map, 9.3)`

```
Save()
saveACopy(filename, {version})
```

Functions

`<function_name>`

e.g. `ExportToPDF(<map_document>, <path and filename of PDF>)`

`ExportToPDF(map, path and PDF file name)` – **exports map document to PDF formatted file**

`ListPrinterNames()` – lists printers on the local machine

`PrintMap(map, {printer name}, ...)` – prints the map document to a printer on the local machine

Data Frame

Properties

`<dataframe_variable>.<property> = <value>`

e.g. `dataframe.scale = 24000`

```
rotation
scale
extent
spatialReference
mapUnits
elementHeight(page units)
elementWidth(page units)
elementPostionX(page units)
elementPositionY(page units)
```

Methods

`<dataframe_variable>.<method>`

e.g. `dataframe.panToExtent()`

`panToExtent` - **extent**
`zoomToSelectedFeatures()`

Functions

e.g. `dataframe = ListDataframes(mxd) [0]`

e.g. `AddLayer(dataframe, aLayer, 'AUTO_ARRANGE')`

`ListDataFrames(map, {wildcard}) [position]` – lists data frames; position is the position in a Python list of dataframes (if only one data frame, then the example is typically used indicating the "first" data frame in the map)

`AddLayer(data frame, layer, {position})` - **adds a layer (from disk or within a map document to a map frame at an optional position)**

`InsertLayer(data frame, reference layer, insert layer, {insert position})` - **inserts a layer from disk or within a map document to a specific location in a data frame with respect to a reference layer and optional position; more precise way of adding a layer to a map document**

`MoveLayer(data frame, reference layer, move layer, {insert position})` - **moves a layer to a different location within a map document based on an existing reference layer and optional position**

`RemoveLayer(data frame, remove layer)` - **removes a layer from the the data frame. If more than one layer has the same name a loop must be used to remove all occurrences of the layer**

`UpdateLayer(data frame, update layer, source layer, {symbology only})` - **update all layer properties (or only symbology) based on a separate source layer's properties. Only symbology can be changed if the optional symbology only parameter is used**

Layer

Properties

\<layer_variable\>.\<property\> = \<value\>
e.g. `aLayer.name == 'City Boundary'`

`dataSource`
`definitionQuery`
`name` – the name of the layer in the Table of Contents
`longName` (to determine if layer is part of a group layer)
`showLabels`
`transparency`
`visible` (True/False)
`workspacePath`

Methods

```
<layer_variable>.<method>
e.g. aLayer.getExtent()

getExtent({symbolized extent})*
getSelectedExtent({symbolized extent})*
```
save() - saves the existing .lyr file (See ArcGIS Help for more details)
`saveACopy(.lyr filename, {version})` – can use this to create a new layer file and not overwrite an existing layer file.

* default is TRUE so that the map extent includes the symbols so they are not cut off in the data frame

Function

```
e.g. LayersList = ListLayers(mxd)

ListLayers(map or layer, {wildcard}, {dataframe})*
```

* Layers can exist in a map document, data frame or layer (.lyr) files

Layout Elements

Properties

See the specific Layout Elements below after the Layout Element function.

Method

```
adjustColumnCount(column count)
```

Function

```
e.g. tElements = ListLayoutElements(mxd, "TEXT_ELEMENT")

ListLayoutElements(map, {element type}, {wildcard})*
```

*Default is no element type. May need a specific element type qualifier such as one of the following:

```
GRAPHIC_ELEMENT
DATAFRAME_ELEMENT
GRAPHIC_ELEMENT
LEGEND_ELEMENT
MAPSURROUND_ELEMENT
PICTURE_ELEMENT
TEXT_ELEMENT
```

The different element type properties are described below.

Graphic Element

e.g. `tElement.elementPositionX = 4.5`

```
elementHeight
elementPositionX
elementPostitionY
elementWidth
name -  name of element, string
type
```

Legend Element (units for elements are in page units)

```
autoAdd -  autoAdd layer to legend
elementHeight
elementPositionX
elementPostitionY
elementWidth
items -  a string list of legend items
name -  name of element, string
parentDataFrameName  - name of Data Frame associated with legend
title -  legend title
type
```

Mapsurround Element (units for elements are in page units)

```
elementHeight
elementPositionX
elementPostitionY
elementWidth
```
name - **name of element, string**
`parentDataFrameName` - **name of Data Frame associated with legend**
`type`

Picture Element (units for elements are in page units)

```
elementHeight
elementPositionX
elementPostitionY
elementWidth
```
name - **name of element, string**
`sourceImage` - **path to image source**
`type`

Text Element (units for elements are in page units)

```
elementHeight
elementPositionX
elementPostitionY
elementWidth
```
name - **name of element, string**
`text` - **text string that show up on the map**
`type`

References

Author's website – www.jenningsplanet.com
Jennings, N. "Managing Street Sign Assets: An enterprise geospatial business systems integration solution." *ArcUser*, Winter 2009. Date Accessed: 11.09.2011
http://www.Esri.com/news/arcuser/0109/streetsigns.html

ArcGIS

ArcGIS Resource Center - http://resources.arcgis.com/
ArcGIS Web-based Help - http://resources.arcgis.com/content/web-based-help
ArcGIS Blog - http://blogs.Esri.com
ArcGIS Forums - http://forums.arcgis.com/forums/117-Python
Geoprocessing script examples and models - http://resources.Esri.com/geoprocessing/index.cfm?fa=codegallery
Esri Training courses - http://training.Esri.com/acb2000/showdetl.cfm?did=6&Product_id=971
ArcUser - http://www.Esri.com/news/arcuser
ArcGIS Resource Center. Exception Code Snippet. Esri, 2011. http://help.arcgis.com/en/arcgisdesktop/10.0/help/index.html#//002z0000000q000000.

Python

Python website – www.python.org
Lutz, Mark. Learning Python, 4th Ed. Beijing: O'Reilly Media, Inc., 2009.

Organizations

Esri – www.Esri.com
American River College GIS Program - http://wserver.arc.losrios.edu/~earthscience/
City of Sacramento GIS – www.cityofsacramento.org/gis
County of Sacramento GIS – www.sacgis.org
Cal Atlas – http://atlas.ca.gov

Index

`.lstrip`, 304
`.path.join`, 289, 291
`.path.splitext`, 291
`.rstrip`, 297
`.sep`, 275, 276, 288, 289, 291, 296, 297
`.strip`, 289

A

`AddError`, 407, 408, 429
`AddField`, 71, 216, 217, 252, 273
`AddIndex`, 267
`AddJoin`, 238, 239, 240, 268, 274
`AddLayer`, 354, 355, 377, 453, 454
`AddMessage`, 405, 406, 407, 408, 419, 429
`AddWarning`, 407, 429
ArcToolbox, 12, 14, 15, 27, 28, 31, 38, 42, 45, 48, 50, 52, 53, 54, 56, 59, 64, 110, 121, 126, 139, 156, 158, 275, 383, 386, 388, 407, 417
attribute index, 234, 238

B

backslash, 88, 89, 112, 170
batch file, 431, 432, 437, 438, 439, 441, 446, 447, 448

C

cast, 83, 180, 195, 316, 318
Check Module, 39, 40, 80, 92, 96, 97, 99, 103, 105, 132, 184, 219
CheckOutExtension, 292, 314
Copy Features, 156, 171, 196
`CreateTable`, 216, 217, 251, 273, 275
cursors, 139, 146, 159, 175, 205, 206, 208, 209, 210, 219, 224, 229, 238, 241, 243, 250, 257, 258, 265, 266, 274, 451
  InsertCursor, 209, 216, 220, 274, 276
  SearchCursor, 21, 89, 90, 91, 92, 209, 210, 212, 225, 274
  UpdateCursor, 209, 223, 225, 226

custom tool, 27, 28, 381, 385, 386, 405, 406, 410, 411, 428

D

data frame, 324, 325, 327, 328, 332, 333, 334, 336, 337, 338, 339, 341, 342, 347, 350, 352, 353, 355, 356, 357, 359, 361, 362, 363, 364, 365, 368, 371, 373, 374, 375, 377, 378, 453, 454, 455
data management, 15, 19, 27, 50, 432, 446
data path, 65, 68, 70, 112, 113, 115, 137, 186, 224, 258, 259, 309
definition query, 162, 347, 348, 349, 373, 374
`definitionQuery`, 347, 348, 373, 454
`del`, 222, 256, 265, 272, 274
`Delete_management`, 128, 156, 275
`Describe`, 137, 278, 281, 282, 283, 290, 291, 292, 293, 294, 300, 302, 304, 378

E

`elementHeight`, 365, 453, 456, 457
`elementPositionX`, 365, 456, 457
`elementPositionY`, 365, 453
`elementWidth`, 365, 453, 456, 457
error handling, 14, 28, 39, 42, 43, 45, 62, 87, 93, 96, 123, 139, 305, 306, 309, 314, 417
error messages, 32, 45, 76, 96, 100, 132, 174, 305, 306, 309, 314, 319, 409
escape characters, 93, 155
except, 39, 42, 43, 44, 45, 62, 79, 86, 87, 96, 99, 102, 103, 105, 123, 129, 152, 178, 179, 185, 186, 245, 250, 258, 295, 305, 306, 308, 309, 314, 399, 408, 409
exception, 42, 99, 129, 130, 179, 186, 203, 306, 307, 308, 315
`Exists`, 128, 137, 174, 178, 182, 198, 203, 252, 274, 275, 289, 291, 297, 300, 311
`ExportToPDF`, 344, 345, 366, 369, 452

F

feature class, 18, 20, 21, 36, 49, 70, 71, 78, 81, 83, 84, 90, 96, 114, 121, 126, 134, 135, 146, 158, 159, 161, 163, 166, 169, 170, 171, 173, 178, 179, 180, 181, 187, 196, 202, 203, 204, 205, 209, 210, 211, 216, 220, 223, 224, 230, 231, 232, 233, 235, 236, 237, 239, 241, 242, 252, 262, 263, 266, 267, 268, 269, 270, 271, 273, 274, 275, 276, 277, 279, 281, 288, 296, 297, 298, 314, 317, 356, 393, 394, 395, 396, 397, 401, 402, 407, 417

feature classes, 15, 27, 59, 78, 81, 83, 86, 110, 113, 121, 145, 146, 156, 159, 160, 170, 182, 198, 210, 218, 229, 236, 238, 243, 244, 250, 259, 266, 281, 286, 287, 288, 289, 296, 297, 324, 354, 395, 396, 423

feature layer, 49, 82, 96, 144, 156, 158, 159, 160, 161, 163, 165, 166, 167, 168, 169, 170, 178, 179, 180, 181, 183, 187, 191, 192, 210, 238, 239, 240, 266, 267, 268, 272, 274

foreground processing, 389

forward slash, 89

G

geodatabase, 21, 49, 71, 83, 90, 113, 114, 147, 152, 156, 210, 216, 217, 224, 226, 227, 242, 250, 257, 258, 259, 261, 262, 263, 264, 266, 273, 274, 275, 276, 281, 286, 287, 288, 289, 296, 297, 298, 393, 397

geoprocessing, 11, 13, 14, 15, 16, 17, 18, 19, 20, 25, 27, 28, 29, 30, 31, 34, 35, 36, 37, 38, 39, 45, 47, 50, 56, 62, 81, 82, 83, 86, 96, 98, 101, 107, 109, 110, 112, 115, 116, 118, 119, 122, 123, 126, 127, 139, 142, 156, 158, 159, 161, 166, 169, 170, 176, 181, 185, 187, 204, 231, 253, 286, 289, 290, 333, 383, 413, 431, 432, 443, 446, 458

`GetCount`, 179, 180, 181, 183, 195, 314, 316, 317

`getExtent`, 348, 349, 374, 379, 455

`GetParameterAsText`, 401, 402, 407, 419, 429, 436, 437, 447

`getSelectedExtent`, 350, 352, 371, 379, 455

`getValue`, 214, 242, 243, 271, 273, 274

group layer, 339, 351, 363, 377, 454

H

hard coded, 115, 116, 137, 187, 242, 269, 289, 297, 337, 345, 401

I

`if` statement, 79, 85, 103, 135, 182, 217, 252, 294, 296, 316, 317

L

layer file, 126, 354, 355, 375, 377, 379, 455

layers, 47, 49, 50, 81, 110, 119, 121, 145, 146, 152, 156, 158, 160, 161, 166, 170, 182, 190, 195, 198, 203, 209, 210, 238, 239, 266, 273, 300, 322, 323, 324, 325, 328, 329, 332, 333, 334, 338, 339, 347, 348, 350, 351, 354, 357, 359, 360, 363, 364, 368, 376

layout elements, 325, 326, 327, 329, 332, 333, 334, 341, 342, 343, 347, 355, 359, 360, 364, 365, 368, 371, 376

line continuation character, 194

`ListDataFrames`, 111, 112, 336, 338, 361, 453

`ListFields`, 84, 240, 241, 268, 274

`ListIndexes`, 267, 274

`ListLayers`, 338, 339, 341, 348, 363, 455

`ListLayoutElements`, 342, 343, 369, 455

`ListLegendEelements`, 375

Lists, 83, 268, 281, 286

log files, 139, 305, 310, 311, 312, 319, 446

`longName`, 339, 340, 363, 454

Loops
  for loop, 79, 208
  while loop, 171, 214, 215, 216, 221, 222, 244, 254, 257

M

Make Feature Layer, 49, 145, 146, 147, 158, 159, 160, 161, 162, 163, 178, 185, 187, 188, 191, 192, 204, 210

Make Table View, 146, 158, 160, 161, 162, 204, 210

map document, 53, 300, 321, 328, 329, 333, 334, 335, 336, 338, 339, 341, 343, 345, 346, 347, 348, 351, 354, 355, 359, 360, 361, 363, 364, 366, 368, 376, 378, 379, 452, 453, 454, 455

map elements, 321, 322, 325, 328, 329, 333, 347, 357, 368
map extent, 347, 348, 349, 455
map template, 328, 329, 346
`MapDocument`, 111, 112, 335, 360
mapping module, 20, 23, 31, 112, 139, 321, 327, 328, 329, 332, 333, 334, 335, 343, 359, 360, 452
ModelBuilder, 12, 15, 17, 20, 25, 34, 47, 48, 49, 50, 51, 52, 56, 59, 61, 107, 158, 383

N

new line feed continuation character, 195

P

Parameter Properties, 391, 393, 394, 398, 419, 420, 421, 429
parameters, 27, 31, 34, 36, 38, 39, 48, 50, 56, 57, 59, 63, 65, 70, 71, 82, 83, 88, 89, 97, 98, 100, 110, 115, 116, 119, 120, 121, 122, 126, 135, 137, 139, 143, 144, 145, 159, 161, 163, 166, 167, 168, 170, 181, 187, 188, 191, 194, 210, 223, 238, 239, 240, 253, 267, 273, 275, 278, 287, 381, 386, 391, 392, 393, 394, 396, 397, 398, 400, 401, 402, 403, 404, 407, 410, 411, 417, 418, 419, 420, 423, 426, 427, 428, 429, 431, 432, 435, 436, 437, 441, 447
print statement, 92, 99, 102, 103, 104, 105, 171, 180, 183, 195, 196, 197, 198, 202, 248, 256, 257, 264, 283, 285, 308, 311, 313, 318
`PrintMap`, 344, 345, 366, 453
pseudo-code, 18, 108, 123, 137, 186, 192, 195
Python
    IDLE, 25, 31, 32, 33, 36, 51, 67, 70, 73, 75, 80, 96, 97, 98, 102, 105, 119, 123, 129, 149, 173, 174, 176, 184, 219, 222, 257, 383, 401, 417, 428, 431, 447
    Python Shell, 32, 34, 35, 36, 44, 45, 61, 73, 75, 76, 77, 92, 98, 100, 104, 105, 127, 132, 133, 135, 138, 166, 171, 174, 176, 180, 184, 195, 214, 219, 222, 240, 241, 248, 249, 268, 270, 271, 273, 282, 302, 310, 312, 317, 318, 335, 340, 361, 362, 405, 435, 447
Python class, 307
Python Constructs
    capitalization, 40, 78, 122
    comments, 80, 81, 95
    indentation, 79, 213, 244, 291
Python function, 93, 95, 256
Python Modules
    arcpy, 29, 30, 31, 35, 61, 81, 84, 86, 87, 88, 89, 110, 111, 112, 113, 121, 123, 124, 128, 131, 137, 161, 178, 186, 220, 223, 245, 250, 258, 275, 276, 282, 286, 287, 289, 292, 297, 300, 317, 334, 335, 360, 401, 402, 405, 436, 437
    datetime, 312, 314, 318, 365
    os, 86, 87, 102, 111, 112, 244, 275, 276, 288, 289, 291, 296, 297
    sys, 86, 87, 111, 112, 124, 130, 131, 138, 178, 185, 186, 245, 250, 258, 292, 402
    time, 313
    traceback, 39, 43, 44, 45, 86, 87, 102, 112, 124, 130, 131, 138, 178, 185, 186, 245, 250, 258, 292, 408

Q

query syntax, 98, 145, 146, 189, 198, 237, 300
querying data, 27, 152
Quotes
    double quotes, 82, 88, 89, 90, 91, 93, 119, 144, 146, 147, 194, 253, 436
    single quotes, 89, 90, 91, 92, 144, 147, 149, 150, 153, 253

R

`raise`, 308, 316
raw string suppression, 89
`RemoveIndex`, 267
`RemoveJoin`, 271, 272, 274

S

`save()`, 346, 354, 355, 379, 455
`saveACopy`, 346, 347, 355, 366, 375, 379, 452, 455
`saveAsCopy`, 354
schema lock, 218, 219
`SelectLayerByAttribute`, 90, 91, 92, 99, 100, 139, 144, 161, 163, 164, 165, 166, 167, 170, 176, 210, 302, 306, 314, 317, 350, 353, 371
`SelectLayerByLocation`, 139, 144, 166, 167, 169, 170, 176, 177, 306, 314, 350, 357
`setValue`, 222, 274, 277

showLabels, 354, 355, 454
Spatial Analyst, 86, 283, 291, 292, 300, 303, 314
strings, 78, 81, 82, 83, 88, 89, 90, 113, 115, 148, 149, 151, 152, 153, 155, 194, 204, 244, 252, 291, 302, 314, 357, 397

T

table joins, 139, 218, 230, 235
table view, 82, 96, 156, 158, 160, 161, 165, 180, 210, 238, 239, 240, 266, 267, 268, 272, 274
tables, 15, 17, 21, 27, 37, 78, 81, 86, 110, 145, 146, 156, 158, 159, 160, 161, 182, 198, 205, 209, 229, 230, 231, 232, 234, 235, 236, 237, 238, 240, 242, 243, 244, 259, 266, 268, 274, 281
text elements, 325, 328, 333, 342, 343, 357, 364, 365, 374
Tool documentation, 410
toolbox alias, 121, 127, 128
try, 44, 48, 62, 79, 87, 96, 99, 102, 103, 105, 123, 127, 132, 148, 178, 179, 185, 186, 188, 192, 196, 203, 226, 305, 306

U

Universal Network Connection, 114

V

variables, 13, 40, 44, 71, 76, 78, 81, 82, 87, 95, 98, 102, 103, 105, 107, 114, 115, 116, 119, 120, 122, 123, 126, 127, 134, 137, 138, 139, 145, 169, 179, 182, 187, 188, 192, 203, 222, 229, 242, 247, 251, 256, 260, 263, 266, 269, 270, 271, 273, 275, 291, 292, 293, 357, 376, 401, 403

W

Windows Scheduler, 431, 439
workspace, 88, 89, 113, 114, 123, 124, 125, 137, 179, 186, 210, 223, 224, 245, 250, 251, 258, 259, 266, 274, 275, 287, 288, 289, 292, 296, 297, 308, 309, 334, 397, 435, 437

Z

zoomToSelectedFeatures, 332, 350, 351, 368, 370, 379, 453

Made in the USA
Lexington, KY
29 August 2012